Lecture Notes in Computer Science 6653

Commenced Publication in 1973
Founding and Former Series Editors:
Gerhard Goos, Juris Hartmanis, and Jan van Leeuwen

Allan Hanbury   Andreas Rauber
Arjen P. de Vries (Eds.)

Multidisciplinary
Information Retrieval

Second Information Retrieval Facility Conference
IRFC 2011
Vienna, Austria, June 6, 2011
Proceedings

 Springer

Volume Editors

Allan Hanbury
Information Retrieval Facility
Donau City Str. 1, 1220 Vienna, Austria
E-mail: a.hanbury@ir-facility.org

Andreas Rauber
Vienna University of Technology
Favoritenstr. 9-11/188, 1040 Vienna, Austria
E-mail: rauber@ifs.tuwien.ac.at

Arjen P. de Vries
Centrum Wiskunde and Informatica
Science Park 123, 1098 XG Amsterdam, The Netherlands
E-mail: arjen@acm.org

ISSN 0302-9743   e-ISSN 1611-3349
ISBN 978-3-642-21352-6   e-ISBN 978-3-642-21353-3
DOI 10.1007/978-3-642-21353-3
Springer Heidelberg Dordrecht London New York

Library of Congress Control Number: Applied for

CR Subject Classification (1998): H.3, I.2.4, H.5, H.4, I.2, C.2

LNCS Sublibrary: SL 3 – Information Systems and Application, incl. Internet/Web
and HCI

Typesetting: Camera-ready by author, data conversion by Scientific Publishing Services, Chennai, India

Printed on acid-free paper

Springer is part of Springer Science+Business Media (www.springer.com)

Preface

These proceedings contain the refereed papers presented at the Second Information Retrieval Facility Conference (IRFC), which was held in Vienna on June 6, 2011. The Second IRFC aimed to tackle four complementary research areas:

– Information retrieval
– Semantic Web technologies for IR
– Natural language processing for IR
– Large-scale or distributed computing for the above areas

IRFC 2011 received 19 high-quality submissions, of which 10 were accepted for presentation at the conference and appear here. Three of these fall into the natural language processing for IR area, one into the distributed computing area, while the remaining six are IR papers.

The IRF Conference provides a multi-disciplinary, scientific forum that aims to bring young researchers into contact with industry at an early stage. The link to industry was emphasized in 2011 with the addition of a panel discussion, where industry representatives considered the practical implications of the work presented.

The Information Retrieval Facility (IRF) is an open IR research institution, managed by a scientific board drawn from a panel of international experts in the field. As a non-profit research institution, the IRF provides services to IR science in the form of a reference laboratory and hardware and software infrastructure. Committed to Open Science concepts, the IRF promotes publication of recent scientific results and newly developed methods, both in traditional paper form and as data sets freely available to IRF members. Such transparency ensures objective evaluation and comparability of results and consequently diversity and sustainability of their further development.

The IRF is unique in providing a powerful supercomputing infrastructure that is exclusively dedicated to semantic processing of text. The first standardized dataset made available by the IRF is MAREC, a collection of 19 million patent documents. These data are housed in an environment that allows large-scale scientific experiments on ways to manage and retrieve this knowledge. Subsets of MAREC are used in the CLEF-IP and TREC-CHEM evaluation campaigns, providing researchers with the opportunity to experiment with realistic retrieval tasks on a real-world data corpus. Two of the papers in this volume make use of the CLEF-IP data.

Our sincere thanks go out to:

– The professional team at the IRF and Vienna University of Technology for their help in preparing the conference and this volume: Marie-Pierre Garnier; Katja Mayer; Mihai Lupu; Veronika Stefanov; Matthias Samwald; Linda Andersson

- The IRF executive board: Francisco Eduardo De Sousa Webber, Daniel Schreiber and Sylvia Thal, for their exceptional support
- Bruce Croft for his support as General Chair of the conference
- The IRF scientific board and John Tait (IRF CSO) for their guidance
- Maarten de Rijke, University of Amsterdam, the Netherlands, for accepting to give a keynote
- The panelists
- The members of the Program Committee and the additional reviewers for their thorough reviews
- The sponsors: Google; CEPIS-EIRSG, the European Information Retrieval Specialist Group
- BCS — The Chartered Institute for IT and the Association for Computing Machinary (ACM) for endorsing the conference
- The Vienna University of Technology and the Austrian Computer Society for their organizational assistance

We hope that you enjoy the results!

June 2011   Allan Hanbury
                                                                      Andreas Rauber
                                                                      Arjen P. de Vries

Organization

Program Committee

Maristella Agosti	University of Padua, Italy
Yannis Avrithis	National Technical University of Athens, Greece
Christos Bouganis	Imperial College London, UK
Pavel Braslavski	Yandex, Russia
Jamie Callan	Carnegie Mellon University, USA
Kilnam Chon	Keio University, Japan
Philipp Cimiano	University of Bielefeld, Germany
Paul Clough	University of Sheffield, UK
Fabio Crestani	University of Lugano, Italy
Hamish Cunningham	University of Sheffield, UK
Gideon Dror	Yahoo! Research
Norbert Fuhr	University of Duisburg-Essen, Germany
Wilfried Gansterer	University of Vienna, Austria
Gregory Grefenstette	Exalead, France
David Hawking	Funnelback, Australia
Noriko Kando	National Institute of Informatics, Japan
Gabriella Kazai	Microsoft Research, UK
Udo Kruschwitz	University of Essex, UK
Marie-Francine Moens	Katholieke Universiteit Leuven, Belgium
Henning Müller	University of Applied Sciences Western Switzerland
Patrick Ruch	University of Applied Sciences Geneva, Switzerland
Stefan Rueger	The Open University, UK
Frank Seinstra	VU University, Amsterdam, The Netherlands
Oren Somekh	Yahoo! Research, Israel
John Tait	Information Retrieval Facility, Austria
Dolf Trieschnigg	University of Twente, The Netherlands
Howard Turtle	Syracuse University, USA
Keith Van Rijsbergen	University of Glasgow, UK
Suzan Verberne	Radboud University Nijmegen, The Netherlands
Thijs Westerveld	Teezir Search Solutions, The Netherlands

Additional Reviewers

Niels Drost	Jacopo Urbani
Jaap Kamps	Timo Van Kessel
Parvaz Mahdabi	

Sponsored by

In cooperation with

Endorsed by

Table of Contents

Adapting Rankers Online

Katja Hofmann, Shimon Whiteson, and Maarten de Rijke

ISLA, University of Amsterdam, Science Park 904, 1098 XH Amsterdam
{K.Hofmann,S.A.Whiteson,deRijke}@uva.nl

At the heart of many effective approaches to the core information retrieval problem—identifying relevant content—lies the following three-fold strategy: obtaining content-based matches, inferring additional ranking criteria and constraints, and combining all of the above so as to arrive at a single ranking of retrieval units.

Over the years, many models have been proposed for content-based matching, with particular attention being paid to estimations of query models and document models. Different task and user scenarios have given rise to the study and use of priors and non-content-based ranking criteria such as freshness, authoritativeness, and credibility. The issue of search result combinations, whether ranked-based, score-based or both, has been a recurring theme for many years. As retrieval systems become more complex, *learning to rank* approaches are being developed to automatically tune the parameters for integrating multiple ways of ranking documents. This is the issue on which we will focus in the talk.

Search engines are typically tuned offline; they are tuned manually or using machine learning methods to fit a specific search environment. These efforts require substantial human resources and are therefore only economical for relatively large groups of users and search environments. More importantly, they are inherently static and disregard the dynamic nature of search environments, where collections change and users acquire knowledge and adapt their search behaviors. Using *online* learning to rank approaches, retrieval systems can learn directly from implicit feedback, while they are running.

The talk will discuss three issues around online learning to rank: balancing exploitation and exploration, gathering data using one pair of rankers and using it to compare another pair of rankers, and the use of rich contextual data.

Balancing exploitation and exploration. In an online setting, algorithms need to both explore new solutions to obtain feedback for effective learning, and exploit what has already been learned to produce results that are acceptable to users. In recent work [1], we have formulated this challenge as an *exploration-exploitation dilemma* and present the first online learning to rank algorithm that works with implicit feedback and balances exploration and exploitation. We leverage existing learning to rank data sets and recently developed click models to evaluate the proposed algorithm. Our results show that finding a balance between exploration and exploitation can substantially improve online retrieval performance, bringing us one step closer to making online learning to rank work in practice.

Generalizing to novel rankers. Implicit feedback, such as users' clicks on documents in a result list, is increasingly being considered as an alternative to explicit relevance judgments. For example, previous work has shown that click data can be used to detect

A. Hanbury, A. Rauber, and A.P. de Vries (Eds.): IRFC 2011, LNCS 6653, pp. 1–2, 2011.

even small differences between rankers, and that it can be used for online learning to rank. Previous methods can identify the better of two rankers with high confidence, but currently the data collected for comparing one pair of rankers cannot be reused for other comparisons. As a result, the number of rankers that can be compared is limited by the amount of use of a search engine. In the talk, ongoing work will be presented on re-using previously collected, historical data by applying importance sampling to compensate for mismatches between the collected data and distributions under the target data. We will show that in this way, rankers can be compared effectively using historical data.

Contextual data. The last part of the talk will be forward-thinking and focus on future work and challenges. Most retrieval systems are integrated in a larger contextual setting, where no item is an island. Events in one document stream are correlated with events in another. Increasingly rich declarative models describe the task, workflow, interaction and organisational structure. Structured knowledge, for instance in the form of linked open data, is available in large quantities to help us inform our retrieval algorithms. How can we use these sources of information in an online setting?

Reference

[1] Hofmann, K., Whiteson, S., de Rijke, M.: Balancing exploration and exploitation in learning to rank online. In: Clough, P., Foley, C., Gurrin, C., Jones, G.J.F., Kraaij, W., Lee, H., Mudoch, V. (eds.) ECIR 2011. LNCS, vol. 6611, pp. 251–263. Springer, Heidelberg (2011)

Building Queries for Prior-Art Search

Parvaz Mahdabi, Mostafa Keikha, Shima Gerani,
Monica Landoni, and Fabio Crestani

Faculty of Informatics, University of Lugano, Lugano, Switzerland
{parvaz.mahdabi,mostafa.keikha,shima.gerani,
monica.landoni,fabio.crestani}@usi.ch

Abstract. Prior-art search is a critical step in the examination procedure of a patent application. This study explores automatic query
generation from patent documents to facilitate the time-consuming and
labor-intensive search for relevant patents. It is essential for this task to
identify discriminative terms in different fields of a query patent, which
enables us to distinguish relevant patents from non-relevant patents. To
this end we investigate the distribution of terms occurring in different
fields of the query patent and compare the distributions with the rest
of the collection using language modeling estimation techniques. We experiment with term weighting based on the Kullback-Leibler divergence
between the query patent and the collection and also with parsimonious
language model estimation. Both of these techniques promote words that
are common in the query patent and are rare in the collection. We also incorporate the classification assigned to patent documents into our model,
to exploit available human judgements in the form of a hierarchical classification. Experimental results show that the retrieval using the generated
queries is effective, particularly in terms of recall, while patent description is shown to be the most useful source for extracting query terms.

1 Introduction

The objective of prior-art search in patent retrieval is identifying all relevant
information which may invalidate the originality of a claim of a patent application. Therefore all patent and non patent literature that have been published
prior to the filing date of a patent application need to be searched. As shown by
Azzopardi et al. the most executed search tasks in the patent domain are nov-
elty and patentability searches [3]. In both of these searches a patent examiner is
required to find all previously published materials on a given topic, as missing a
single relevant document can lead to lawsuits due to patent infringement. Thus
patent retrieval is considered a recall-oriented application.

There are several properties which make prior-art search a challenging task,
with different problems and challenges when compared to other search tasks, such
as the web search. The first property is that the starting point of the prior-art task
is the patent document in question. Since in this task the information need is presented by a patent document rather than a short query, a major challenge is how

A. Hanbury, A. Rauber, and A.P. de Vries (Eds.): IRFC 2011, LNCS 6653, pp. 3–15, 2011.

to transform the patent application into search queries [10,27,23]. A variety of different techniques have been employed in previous studies for identifying effective query terms, mainly looking into the distribution of term frequency.

A second property is related to the vocabulary usage in patent domain which is very unique and far from everyday speech and writing and often contains highly specialized or technical words not found in everyday language [2]. Writers tend to purposely use many vague terms and expressions along with non-standard terminology in order to avoid narrowing the scope of their claims [5]. They also develop their own terminologies to increase their chance of acceptance. This frequent intentional obfuscation of content by patent writers results in vocabulary mismatch. For example one patent document may contain few or no keywords in the query topic, but the idea conveyed in the patent maybe quite similar or even identical to the query topic [2].

The last property is linked to the structure of patents. Patent documents are structured documents with different fields such as abstract, description, and claims. Patent Writers use different style of writing for describing the invention in different fields of patent. For example, the abstract and description use a technical terminology while the claims uses a legal jargon [27].

In this paper we explore generating queries from different fields of the patent documents. Our contribution is to build an effective term selection and weighting technique using a weighted log-likelihood based approach to distinguish words which are both indicative of the topic of the query and are not extensively used in the collection. We also investigate query modeling based on the Parsimonious language model for building the topic of the query patent. Furthermore, we utilize the knowledge embedded in IPC classes. This addresses the vocabulary mismatch as we include words in the query which are not present in the query topic itself.

The rest of the paper is structured as follows. We first explain the CLEP-IP 2010 collection and recent work on patent retrieval. We then present an overview of our approach. We define the query generation problem and describe three approaches to estimate the topic of a query patent. Finally, we describe an empirical comparison performed between different query modeling methods for the prior-art task of CLEF-IP 2010.

2   CLEF-IP 2010 Collection

The patent collection released for the prior-art search task of CLEF-IP 2010 constitutes 1.3 million patent documents from the European Patent Office (EPO). The collection is multilingual in nature, and patent documents can be written in English, French or German. Each patent application has one or more International Patent Classification (IPC) classes assigned to it. IPC exhibits a hierarchical order consisting of more than 70,000 subdivisions. This classifications gives a broad technological description of the invention. These assignments are performed by patent examiners and are used by all patent offices [14]. Patent documents have multiple versions which correspond to the different stages of a patent's life cycle [23]. In the relevance judgements released for the CLEF-IP 2010 prior-art task, different versions of a patent are expected to be found.

The structure of a prior-art task topic is as follows:

```
<topic >
<num>PAC-number</num>
<narr>Find all patents in the collection that potentially inva-
lidate patent application patentNumber. </narr>
<file>fileName.xml </file>
</topic>
```

As mentioned before, the information need is represented by a document rather than a query, so participants have to first generate a query from the patent document.

The CLEF-IP training set contains documents and relevance judgements for 300 topics. The test set consists of two sub-sets, one with 500 topics, and the other with 2000 topics, referred to as small and large test set. We performed our experiments on the english sub-set of the collection and on the large topic set.

3   Related Work

Recently, patent processing has attracted considerable attention in academic research communities, in particular by information retrieval and natural language processing researchers [8].

The main research in patent retrieval started after the third NTCIR workshop [12], where a few patent test collections were released. Starting from the fourth NTCIR [7], a search task related to the prior-art search was presented which was referred to as *invalidity search run*. The goal was to find prior-art before the filing date of the application in question which conflicts with the claimed invention. The citation parts of the applications are removed and counted as relevant documents used for evaluation of results. Participants performed different term weighting methods for query generation from the claims. They applied query expansion techniques by extracting effective and concrete terms from description to enhance the initial query.

In [25] authors study the rhetorical structure of a claim. They segmented a claim into multiple components, each of which is used to produce an initial query. They then searched for candidate documents on a component by component basis. The final result was produced from the candidate documents. Similar work has been introduced in [19] where the authors analyze the claims structure to enhance the retrieval effectiveness. A claim structure usually consists of *premise* and *invention* parts, which describe existing and new technologies, respectively. Authors proposed a two stage process where they first extract query terms from premise part to increase the recall. They then aim at increasing the precision by extracting another query from invention part. The final relevance score for each document is calculated by merging the scores of the two stages.

IPC classification has been used as an extra feature beside the content of the patent. Different methods for combining text content and classification were proposed. In [25] the authors use IPC codes for document filtering and show

how this feature can help in patent retrieval. In [9] authors integrate IPC codes with a probabilistic retrieval model. They employ the IPC codes for estimating the document prior.

In [6], Fujii applied link analysis techniques to the citation structure of patents. They calculate two different scores based on textual information and citation information. They showed that by combining these two scores better performance can be achieved.

CLEF-IP is another important evaluation platform for comparing performance of patent retrieval systems. CLEF-IP has been running since 2009 and a variety of different techniques have been employed for identifying effective query terms mainly looking into distribution of term frequency [22].

Lopez et al. [14] constructed a restricted initial working set by exploiting the citation structure and IPC metadata. In [15] Magdy et al. generate the query out of the most frequent unigrams and bigrams. In this work the effect of using bigrams in query generation was studied but the retrieval improvement was not significant. This is perhaps because of the unusual vocabulary usage in patent domain.

Graf et al. [10] study the integration of knowledge representations into patent retrieval. They aim to address the vocabulary gap and enable identification of similar documents to a query even in the absence of mutually shared terms. To this cause they model the knowledge as a hierarchical conceptual structure extracted from available IPC information. They extracted representative vocabularies for the technological aspects covered by specific IPC elements.

Xue et al. [27] combined different type of search features with a learning to rank method for transforming a patent to query terms. They explored different factors for a successful transformation such as the structural information of a patent and use of noun phrases besides unigrams.

Bashir et al. [5] analyze the retrievability of patent documents using retrievability measurements [4]. They report a large subset of patents has very low retrievability scores. In order to increase the retrievability of the low retrievable subset, they perform query expansion with Pseudo Relevance Feedback (PRF). They first select a set of relevant terms to improve retrievability by considering the proximity distribution of those relevant terms with respect to the original query terms. They then identify PRF documents based on their similarity with query patents according to these selected terms.

Unlike other IR tasks where the focus is on achieving high retrieval precision, patent retrieval is mostly a recall focused task [16]. On the one hand, the standard mean average precision (MAP) is not suitable for patent search environments because it does not reflect system recall. On the other hand, using the standard recall measure overlooks the user search effort. In [17] authors proposed an evaluation metric, called Patent Retrieval Evaluation Score (PRES), which reflects both the system recall and the user search effort. This metric considers recall and the user search effort. Therefore, this metric is more suitable for recall-oriented applications. In practice, this evaluation metric is more robust when the relevance judgements are incomplete [16].

4   System Architecture

The retrieval system starts with a query patent document, which we aim to find relevant documents for, and generates a ranked list of patent documents. Figure 1 shows the overall architecture of our system for prior-art search. In the first step we generate a query from the patent document. In the second step we formulate the query by selecting the top k terms from the term distribution of the query model. In the third step we retrieve documents relevant to the generated query. We then filter this ranked list by excluding documents which do not have an IPC class in common with the query patent document.

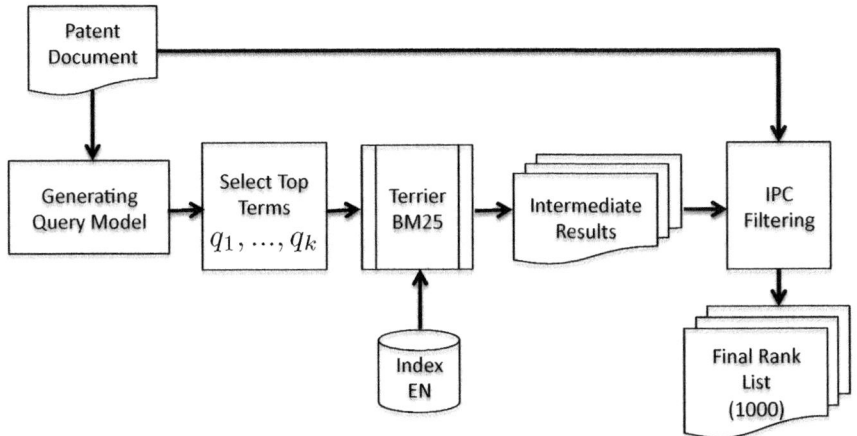

Fig. 1. Overall architecture of the proposed system

In the next section we focus on the query generation problem and propose three methods for performing this step. We limit our experiments to the English subset of the collection. We do not take advantage of citation information in order to determine what can be gained from using only the text information of patent documents.

5   Query Generation for a Query Patent Document

The *query generation for prior-art search* problem is defined as developing an effective algorithm that selects the best terms from the query patent document to form an effective query. An effective query is defined as a query that can better distinguish relevant patents from non relevant patents. We implemented three different approaches to estimate the query model of a patent document. The first two approaches are based on weighted log-likelihood [20], and the third approach is based on parsimonious language modeling [11]. The goal of these approaches are to select the most informative terms for representing the topic of the query patent. These approaches will be discussed in more detail below.

We utilize the structural information of a patent document in our model by building a query model for each field separately. A patent document in the CLEF-IP 2010 collection contains the following fields: the title (*ttl*), the abstract (*abs*), the description (*desc*), and the claims (*clm*). Our aim is to investigate and compare the quality of the extracted terms according to the query model of each field. In an attempt to take into account the full structure of the document, we also explore merging ranked lists generated from different fields.

5.1   Query Model Based on Weighted Log-Likelihood

In our first approach we build a query model (denoted θ_{Q_f}) for the field f of the patent document, where f belongs to {*title, abstract, description, claims*}. We estimate the query model θ_{Q_f} by calculating the relative frequencies for terms in the filed f of the query document. To create a better representation, we smooth the θ_{Q_f} estimate with the topic model of the relevant cluster of documents. This cluster consists of documents with at least one IPC class in common with the query document (denoted $RIPC$). The intuition is that patent documents with similar IPC classes are assumed to have similar topics [10]. This smoothing of the parameters away from their maximum likelihood estimates helps us to exploit the knowledge embedded in the IPC hierarchy in our model. Because of the smoothing, non zero probabilities are assigned to words which are associated with the topic of a document and are not mentioned in the document itself. This can be seen as expanding the document model with the IPC metadata. The query model θ_{Q_f} is estimated as follows:

$$P(w|\theta_{Q_f}) = \lambda \frac{tf(w, Q_f)}{|Q_f|} + \frac{(1-\lambda)}{N} \sum_{d \in RIPC} \frac{tf(w, D_f)}{|D_f|} \qquad (1)$$

$tf(w, Q_f)$ denotes the term frequency of the word w in the field f of the patent document, $|Q_f|$ is the length of the filed f of the patent document, N denotes the size of the relevant cluster $RIPC$, and λ denotes the smoothing parameter. In order to estimate a query model for the patent in question it is necessary to highlight words from the term distribution of θ_{Q_f} which are rare in the collection. To this end, we weight term probabilities in θ_{Q_f} with the following formula:

$$P(w|LLQM_f) \propto p(w|\theta_{Q_f}) log \frac{p(w|\theta_{Q_f})}{p(w|\theta_{C_f})} \qquad (2)$$

where $P(w|\theta_{C_f})$ shows the probability of a word in the collection and is estimated as follows:

$$P(w|\theta_{C_f}) = \frac{tf(w, C_f)}{\sum_{D \in C} |D_f|} \qquad (3)$$

$tf(w, C_f)$ denotes the collection term frequency for the field f. We refer to this model as the Log-Likelihood Query Model ($LLQM_f$). This is a slight variation of the standard weighted log-likelihood ratio [20]. The value in Equation 2 is normalized by a constant, the Kullback-Leibler divergence [18] between the query document Q and the collection language model C. This measure quantifies the similarity of the query document with the topical model of relevance and the dissimilarity between the query document and the collection model. Terms with high divergence are good indicators of the patent document and show the specific terminology of the patent document.

In the second approach, we incorporate the knowledge of the hierarchical classifications of IPC into our model. We estimate a slightly different formulation of the query model, referred to as Cluster Based Query Modeling ($CBQM_f$), by weighting term probabilities in θ_{Q_f} by their relative information in the cluster language model θ_{Cl_f} and the collection language model θ_{C_f}. This model assigns a high score to query terms which are similar to the cluster language model but dissimilar to the collection language model. We base this estimate on the divergence between θ_{Q_f} and the cluster language model, measuring this divergence by determining the log-likelihood ratio between θ_{Q_f} and θ_{Cl_f}, normalized by θ_{C_f}. This formulation gives another way of constructing query model based on the relevant cluster derived from IPC classes.

$$P(w|CBQM_f) \propto p(w|\theta_{Q_f})log\frac{p(w|\theta_{Cl_f})}{p(w|\theta_{C_f})} \tag{4}$$

5.2 Parsimonious Query Modeling

In our third approach we estimate a query model that differentiates the language used by the query patent from the collection model. As suggested by Hiemstra et al. [11], we estimate the topic of the query patent using parsimonious language modeling, by concentrating the probability mass on terms that are indicative of the topic of the query patent but are dissimilar from the collection model. We use Expectation-Maximization (EM) algorithm for estimating the query model of different fields of a patent document. The Parsimonious Query Model (PQM_f) is estimated according to the following iterative algorithm:

E-step:

$$e_w = tf(w, Q_f) \cdot \frac{\lambda P(w|PQM_f)}{(1-\lambda)P(w|C_f) + \lambda P(w|PQM_f)} \tag{5}$$

M-step:

$$P(w|PQM_f) = \frac{e_w}{\sum_w e_w}, \text{ i.e. normalize the model} \tag{6}$$

where $P(w|C_f)$ is the maximum likelihood estimate for the collection and is calculated according to Equation 3. The initial value for $P(w|PQM_f)$ is based on the maximum likelihood estimate for the query as in Equation 1, skipping the smoothing step. The advantage of this estimation model is that it discards field-specific stop-words automatically. This is because we estimate the query model for each field separately. For example, for the abstract field the set of words "system", "device", "apparatus", and "invention" are identified as stop-words.

6   Experimental Methodology and Results

In this section we first explain our experimental setup for evaluating the effectiveness of our proposed methods. We then explain the experiments that we conducted in order to evaluate the effectiveness of different setting of the proposed methods in section 5.

6.1   Experimental Setup

We index the collection using Terrier[1]. Our pre-processing is minimal and involves stop-word removal and stemming using the Porter stemmer. For all our retrieval experiments we use the Terrier implementation of BM25. In the experiments, we compare our term selection techniques with participants of CLEF-IP 2010 [21].

6.2   Parameter Settings

We select the top k terms from the generated query models and submit them as a weighted query to Terrier. The retrieval function is BM25. We then filter the retrieved results based on IPC classes. The proposed models have two parameters: the field f of query patent used for building the query model, and the query length parameter k which denotes the maximum number of selected terms to be used. We first tune these parameters on the training set. Note that the range of the query length parameter for different fields is different. The smoothing parameter λ in $LLQM_f$ and PQM_f is experimentally set to 0.9.

6.3   Effect of Query Length and Field

Tables 1 – 4 show the result of selecting different number of terms from different fields of the query document using the three query estimation approaches introduced in the previous sections. Results are reported over the training set. Note that the query model is built for each field separately. However, in the retrieval step, all the fields of the patent documents are considered for similarity score calculation. In other words, the query model built from field f is applied on all the fields of the patent documents. In the tables, for the sake of readability, the field f used for query estimation is denoted within the parentheses.

The results of all four tables show that increasing the query length improves the evaluation scores. However, when the query length exceeds a limit, adding more candidate query terms does not further improve the performance. This is true for all the three query estimation methods. Based on these experiments we limit the length of the generated queries from description, claims, abstract, and title to 100, 100, 50, 10, respectively.

Table 5 reports the performance of the three term selection techniques on the training set over different fields, using the optimized query length. Furthermore, Table 5 shows the effect of merging multiple search results of the different algorithms using CombSUM and CombMNZ [24]. CombSUM combination method

---

[1] http://terrier.org/

Table 1. Evaluation scores of the different query estimation methods using the description field on the training set for the English subset

PQM(desc)	25	50	75	100	125	150
MAP	0.08	0.09	0.09	**0.10**	0.10	0.09
Recall	0.56	0.57	0.59	**0.59**	0.58	0.57
$CBQM$(desc)	25	50	75	100	125	150
MAP	0.08	0.09	0.10	**0.11**	0.10	0.09
Recall	0.58	0.59	0.60	**0.60**	0.59	0.59
$LLQM$(desc)	25	50	75	100	125	150
MAP	0.08	0.08	0.11	**0.12**	0.12	0.11
Recall	0.59	0.62	0.62	**0.63**	0.61	0.60

Table 2. Evaluation scores of the different query estimation methods using the claims field on the training set for the English subset

PQM(clm)	25	50	75	100	125	150
MAP	0.04	0.05	0.06	**0.07**	0.07	0.07
Recall	0.48	0.50	0.52	**0.54**	0.53	0.52
$CBQM$(clm)	25	50	75	100	125	150
MAP	0.05	0.06	0.06	**0.07**	0.06	0.06
Recall	0.49	0.52	0.53	**0.56**	0.54	0.52
$LLQM$(clm)	25	50	75	100	125	150
MAP	0.06	0.08	0.10	**0.10**	0.09	0.09
Recall	0.51	0.53	0.56	**0.57**	0.56	0.55

calculates the summation of the set of relative ranks, or similarity values, retrieved by multiple search runs. CombMNZ, performs similar to CombSUM by calculating the numerical mean of the set of similarity values and it also provides higher weights to documents retrieved by multiple retrieval methods.

Experiments show that extracting terms from the description field has the best performance over all other fields. The reason for this is the technical language used in description as opposed to the legal jargon which is the characteristic of the claims field. We believe the short length of titles is the reason why selecting terms from the title performs poorly when compared to other fields. Prior work [28] suggests that both the abstract and description use technical terminology, but our results show using the abstract field to be less effective. Further investigation is needed to understand why query terms extracted from the abstract field are not as effective as the those extracted from the description.

Another observation is that $LLQM_f$ outperforms $CBQM_f$ and PQM_f in terms of both MAP and Recall. The reason that $CBQM_f$ performed slightly worse than $LLQM_f$, is perhaps due to the fact that in $CBQM_f$ we consider all documents which have IPC classes in common with the query as feedback documents. This generated cluster of relevant documents is very big, therefore we loose the specific terms which are representative of the topic of the query document.

Table 3. Evaluation scores of the different query estimation methods using the abstract field on the training set for the English subset

PQM(abs)	10	20	30	40	50
MAP	0.05	0.05	0.06	0.06	**0.07**
Recall	0.47	0.48	0.50	0.52	**0.54**
$CBQM$(abs)	10	20	30	40	50
MAP	0.05	0.05	0.06	0.06	**0.07**
Recall	0.48	0.52	0.54	0.56	**0.56**
$LLQM$(abs)	10	20	30	40	50
MAP	0.05	0.05	0.06	0.07	**0.07**
Recall	0.50	0.52	0.54	0.55	**0.56**

Table 4. Evaluation scores of the different query estimation methods using the title field on the training set for the English subset

PQM(tit)	5	10
MAP	0.03	**0.03**
Recall	0.48	**0.50**
$CBQM$(tit)	5	10
MAP	0.04	**0.04**
Recall	0.52	**0.53**
$LLQM$(tit)	5	10
MAP	0.04	**0.05**
Recall	0.52	**0.53**

Our attempt to merge results of the different fields using CombSUM and CombMNZ did not improve the performance of the best setting. Similar results were found when building a single query by combining the selected query terms from different fields, therefore we did not report the results.

6.4   Comparison with the CLEF-IP 2010 Participants

We fix our two parameters for the estimation method of query model, namely the query length and the query field, to the values which have been shown to achieve the best performance on the training set. Now we present our results following this setting on the test set. Our results on the training set show that $LLQM_f$ and $CBQM_f$ perform better than PQM_f. Thus we only present the results of these two approaches on the test set. If we would have submitted the results of $LLQM_f$ approach, it would have ended up on the top-3 for the prior-art task in terms of Recall and PRES. In terms of MAP it would have been placed at rank 4, while $CBQM_f$ would have been placed two ranks below $LLQM_f$.

Table 6 shows our position with respect to other CLE-IP 2010 participants. In our techniques, we did not look into citations proposed by applicants. Among the top ranked participants, only two other approaches by Magdy and Jones [15] and Alink et al. [1] were similar to us in this respect, which are indicated

Table 5. Comparison of performance of the different query estimation methods using the different fields of a patent document

Run	MAP	Recall
PQM(tit)	0.03	0.50
PQM(abs)	0.07	0.54
PQM(desc)	**0.10**	**0.59**
PQM(clm)	0.07	0.54
CombSUM(all)	0.05	0.55
CombMNZ(all)	0.04	0.54
$CBQM$(tit)	0.04	0.53
$CBQM$(abs)	0.07	0.56
$CBQM$(desc)	**0.11**	**0.60**
$CBQM$(clm)	0.07	0.56
CombSUM(all)	0.09	0.57
CombMNZ(all)	0.07	0.56
$LLQM$(tit)	0.05	0.53
$LLQM$(abs)	0.07	0.56
$LLQM$(desc)	**0.12**	**0.63**
$LLQM$(clm)	0.10	0.57
CombSUM(all)	0.09	0.57
CombMNZ(all)	0.08	0.56

Table 6. Prior-art results for best runs in CLEF-IP 2010, ranked by PRES, using the large topic set for the English subset

Run	MAP	Recall	PRES
humb[14]	0.2264	0.6946	0.6149
dcu-wc[15]	0.1807	0.616	0.5167
LLQM	**0.124**	**0.60**	**0.485**
dcu-nc[15]	0.1386	0.5886	0.483
CBQM	**0.124**	**0.589**	**0.477**
spq[1]	0.1108	0.5762	0.4626
bibtem[26]	0.1226	0.4869	0.3187

by *dcu-nc* and *spq*, respectively. Using the citations is the main reason behind the strong performance of the two top ranked approaches in Table 6. Our two approaches are shown in bold face.

Although previous works [13,25] mainly use claims for query formulation, our results suggest that building queries from the description field can be more useful. This result is in agreement with [28], in which query generation for US patents were explored, and the background summary of the patent was shown to be the best source for extracting terms. Since the background summary in US patents uses technical terminology for explaining the invention, it is considered equivalent to the description field in European patents.

7   Discussion and Future Work

Prior-art task is one of the most performed search tasks in patent domain. The information need for this task is presented by a query document. Converting the document into effective search queries is necessary. In this work, we presented three query modeling methods for estimating the topic of the patent application. We integrate the structural information of a patent document and IPC classification into our model. Our results suggest that description is the best field for extracting terms for building queries. Based on our experiments, combining different fields in query formulation, or merging the results afterwards, is not useful. In the future work, we plan to explore the advantage of using the citation structure and noun phrases in the proposed framework. Using a smaller cluster of similar IPC classes for estimating the topical model should also be explored, in an attempt to avoid adding general terms to the query and selecting more specific terms.

Acknowledgements

Authors would like to thank Information Retrieval Facility (IRF) for supporting this work. We would like to also thank Mark Carman for helpful discussions and valuable suggestions.

References

1. Alink, W., Cornacchia, R., de Vries, A.P.: Building strategies, a year later. In: Workshop of the Cross-Language Evaluation Forum, LABs and Workshops, Notebook Papers (2010)
2. Atkinson, K.H.: Toward a more rational patent search paradigm. In: Proceedings of the 1st ACM Workshop on Patent Information Retrieval, pp. 37–40 (2008)
3. Azzopardi, L., Vanderbauwhede, W., Joho, H.: Search system requirements of patent analysts. In: International ACM SIGIR Conference on Research and Development in Information Retrieval, pp. 775–776 (2010)
4. Azzopardi, L., Vinay, V.: Retrievability: an evaluation measure for higher order information access tasks. In: ACM Conference on Information and Knowledge Management, pp. 561–570 (2008)
5. Bashir, S., Rauber, A.: Improving Retrievability of Patents in Prior-Art Search. In: European Conference on Information Retrieval, pp. 457–470 (2010)
6. Fujii, A.: Enhancing patent retrieval by citation analysis. In: International ACM SIGIR Conference on Research and Development in Information Retrieval, pp. 793–794 (2007)
7. Fujii, A., Iwayama, M., Kando, N.: Overview of Patent Retrieval Task at NTCIR-4. In: Proceedings of NTCIR-4 Workshop (2004)
8. Fujii, A., Iwayama, M., Kando, N.: Introduction to the special issue on patent processing. Information Processing and Management 43(5), 1149–1153 (2007)
9. Fujita, S.: Revisiting the Document Length Hypotheses- NTCIR-4 CLIR and Patent Experiments at Patolis. In: Proceedings of NTCIR-4 Workshop (2004)
10. Graf, E., Frommholz, I., Lalmas, M., van Rijsbergen, K.: Knowledge modeling in prior art search. In: First Information Retrieval Facility Conference on Advances in Multidisciplinary Retrieval, pp. 31–46 (2010)

11. Hiemstra, D., Robertson, S.E., Zaragoza, H.: Parsimonious language models for information retrieval. In: International ACM SIGIR Conference on Research and Development in Information Retrieval, pp. 178–185 (2004)
12. Iwayama, M., Fujii, A., Kando, N., Takano, A.: Overview of patent retrieval task at NTCIR-3. In: Proceedings of NTCIR Workshop (2002)
13. Konishi, K.: Query terms extraction from patent document for invalidity search. In: Proc. of NTCIR 2005 (2005)
14. Lopez, P., Romary, L.: Experiments with citation mining and key-term extraction for prior art search. In: Workshop of the Cross-Language Evaluation Forum, LABs and Workshops, Notebook Papers (2010)
15. Magdy, W., Jones, G.J.F.: Applying the KISS Principle for the CLEF-IP 2010 Prior Art Candidate Patent Search Task. In: Workshop of the Cross-Language Evaluation Forum, LABs and Workshops, Notebook Papers (2010)
16. Magdy, W., Jones, G.J.F.: Examining the robustness of evaluation metrics for patent retrieval with incomplete relevance judgements. In: Multilingual and Multimodal Information Access Evaluation, International Conference of the Cross-Language Evaluation Forum, pp. 82–93 (2010)
17. Magdy, W., Jones, G.J.F.: PRES: a score metric for evaluating recall-oriented information retrieval applications. In: International ACM SIGIR Conference on Research and Development in Information Retrieval, pp. 611–618 (2010)
18. Manning, C., Schütze, H.: Foundations of Statistical Natural Language Processing. MIT Press, Cambridge (1999)
19. Mase, H., Matsubayashi, T., Ogawa, Y., Iwayama, M., Oshio, T.: Proposal of two-stage patent retrieval method considering the claim structure. ACM Transactions on Asian Language Information Processing 4(2), 190–206 (2005)
20. Meij, E., Weerkamp, W., de Rijke, M.: A query model based on normalized log-likelihood. In: ACM Conference on Information and Knowledge Management, pp. 1903–1906 (2009)
21. Piori, F.: CLEF-IP 2010: Prior-Art Candidate Search Evaluation Summary. In: Workshop of the Cross-Language Evaluation Forum, LABs and Workshops, Notebook Papers (2010)
22. Piroi, F.: CLEF-IP 2010: Retrieval Experiments in the Intellectual Property Domain. In: Workshop of the Cross-Language Evaluation Forum, LABs and Workshops, Notebook Papers (2010)
23. Roda, G., Tait, J., Piroi, F., Zenz, V.: CLEF-IP 2009: Retrieval Experiments in the Intellectual Property Domain (2009)
24. Shaw, J.A., Fox, E.A.: Combination of multiple searches. In: TREC 1994 (1994)
25. Takaki, T., Fujii, A., Ishikawa, T.: Associative document retrieval by query subtopic analysis and its application to invalidity patent search. In: ACM Conference on Information and Knowledge Management, pp. 399–405 (2004)
26. Teodoro, D., Gobeill, J., Pasche, E., Vishnyakova, D., Ruch, P., Lovis, C.: Automatic Prior Art Searching and Patent Encoding at CLEF-IP 2010. In: Workshop of the Cross-Language Evaluation Forum, LABs and Workshops, Notebook Papers (2010)
27. Xue, X., Croft, W.B.: Automatic query generation for patent search. In: ACM Conference on Information and Knowledge Management, pp. 2037–2040 (2009)
28. Xue, X., Croft, W.B.: Transforming patents into prior-art queries. In: International ACM SIGIR Conference on Research and Development in Information Retrieval, pp. 808–809 (2009)

Expanding Queries with Term and Phrase Translations in Patent Retrieval

Charles Jochim, Christina Lioma, and Hinrich Schütze

Institute for Natural Language Processing,
Computer Science, Stuttgart University
70174 Stuttgart, Germany
{jochimcs,liomca}@ims.uni-stuttgart.de

Abstract. Patent retrieval is a branch of Information Retrieval (IR) that aims to enable the challenging task of retrieving highly technical and often complicated patents. Typically, patent granting bodies translate patents into several major foreign languages, so that language boundaries do not hinder their accessibility. Given such multilingual patent collections, we posit that the patent translations can be exploited for facilitating patent retrieval.

Specifically, we focus on the translation of patent queries from German and French, the morphology of which poses an extra challenge to retrieval. We compare two translation approaches that expand the query with (i) translated terms and (ii) translated phrases. Experimental evaluation on a standard CLEF-IP European Patent Office dataset reveals a novel finding: phrase translation may be more suited to French, and term translation may be more suited to German. We trace this finding to language morphology, and we conclude that tailoring the query translation per language can lead to improved results in patent retrieval.

Keywords: patent retrieval, cross-language information retrieval, query translation, statistical machine translation, relevance feedback, query expansion.

1 Introduction

Information retrieval (IR) systems used in the domain of patents need to address the difficult task of retrieving relevant yet very technical and highly specific content [20,21]. On one hand, patent content is notoriously difficult to process [1]. This difficulty is often exacerbated by the patent authors themselves who intentionally make their patents difficult to retrieve [2]. On the other hand, patent searchers require an exhaustive knowledge of all related and relevant patents, because overlooking a single valid patent potentially has detrimental and expensive consequences, e.g., infringement and litigation. In practice, this means that on one hand we have a very hard retrieval task, and on the other hand, we have demands for very high retrieval effectiveness. We tackle this difficult problem, by focusing on the multilingual aspect of patents. Since patents are partially translated into one or more languages, a collection of patents can be seen as a

A. Hanbury, A. Rauber, and A.P. de Vries (Eds.): IRFC 2011, LNCS 6653, pp. 16–29, 2011.
© Springer-Verlag Berlin Heidelberg 2011

multilingual corpus. Given such a multilingual patent collection, we posit that one way to improve retrieval is by turning monolingual queries into multilingual queries, hence potentially improving their coverage.

We create multilingual patent queries by query translation. We present two alternatives for this approach: (i) term-by-term translations and (ii) translations that can also involve phrases. In the latter case, there are four different types of translations that can occur in the query:

- term to term translation
- term to phrase translation
- phrase to term translation
- phrase to phrase translation

For brevity, we simply refer to these two approaches as *term translation* and *phrase translation* – even though phrase translations are in reality a superset of term translations.

Our query translation is realized using a domain-specific translation dictionary of terms and phrases. We extract this dictionary from the patent collection used for retrieval, using parallel translations in the patents. Specifically, we identify such parallel translations, align them, and compute the translation probabilities between terms and phrases in the aligned translations. These translations constitute the entries in our domain-specific patent translation dictionary.

Our approach differs from previous work in that we derive a bilingual term and phrase dictionary from the retrieval collection itself – that is, we do not derive the dictionary from unrelated parallel corpora. This aspect of our work is important because it is difficult to obtain good translation coverage when using a generic dictionary or parallel text from a different corpus.

To evaluate our query translation hypothesis, we compare retrieval performance of our multilingual queries versus monolingual queries, using a competitive retrieval model. We further include runs where translation has been realised with Google's competitive MT system [12], *Google Translate*[1], so that our translation approach can be compared to a state of the art and freely available competitive approach. In addition, because our query translation can also be seen as a form of query expansion (since queries are expanded with their translations), we conduct experiments with pseudo-relevance feedback. Experimental evaluation on a standard CLEF-IP [19] dataset, focusing on the morphologically difficult cases of German and French queries, shows good results: Regarding the choice of term versus phrase translation, phrase translation seems to be more beneficial for French than for German. The improvement of our query translation approach is especially beneficial to queries of very poor baseline recall. We also find that our translation approaches are compatible with relevance feedback, and even enhance its performance (when combined with it).

The remainder of this paper is organized as follows. Section 2 overviews related work on patent IR with a focus on translation. Section 3 presents our methodology for translating patent queries. Section 4 describes the experimental evaluation of our approach. Finally, section 5 summarizes this work.

---

[1] http://translate.google.com/

2   Related Work

Thorough patent retrieval includes searching over multiple patent databases and potentially over multiple languages. Due to the multilinguality of the task, advances in cross-lingual IR (CLIR) have been explored to improve patent retrieval, for instance the NTCIR initiative on patent IR and machine translation (MT) [9], and the intellectual property (IP) track of the Cross-Language Evaluation Forum (CLEF) [19].

More generally, in the area of CLIR, translation can be broadly realized using a combination of bilingual dictionaries and/or parallel corpora and/or MT (see [17] for an overview). All three of these resources are covered in this work: we present a way of extracting a bilingual dictionary from a parallel corpus, and we also include experiments where translation is realised using Google's competitive MT system, *Google Translate.*

An early, well-cited phrase-based CLIR study was by Ballesteros and Croft [3], who expanded bilingual dictionaries with phrases and used them effectively in IR. Their definition of a phrase differs from the definition of phrase we use in this paper: They defined phrases grammatically as sequences of nouns and adjective-noun pairs, meaning that their approach required some sort of part-of-speech preprocessing. In this work, we define phrases statistically as any string of words, meaning that no grammatical preprocessing is required. Further studies followed in the 1990s and early 2000s, aiming to improve CLIR performance by first improving translation accuracy. However, more recent studies have shown that even though translation accuracy clearly affects CLIR [12], good IR performance may still be obtainable with suboptimal translation accuracy. For example, Gao et al. [10] obtain better results by using cross-lingual query suggestion than by traditional query translation. Combining different monolingual and bilingual resources, they develop a discriminative model for learning cross-lingual query similarity. With this model, they find target queries in a collection of query logs, which are similar to, but not direct translations of, the source queries. Another example to depart from exact query translation, is the work of Wang and Oard [22], who tackle translation for IR as a case of *meaning matching*. Specifically, they align candidate term translations from parallel corpora, which they then augment with WordNet synset (i.e. meaning matching) information. Motivated by these more recent advances into CLIR, in this work we also adopt a translation approach that does not aim to translate the query as accurately as possible, but rather to "gist" the query and translate its most salient parts – to our knowledge, this is novel for patent IR.

Our approach of translating queries consists in expanding them with their respective translations, hence it may be seen as a form of query expansion. Query expansion in general has been shown to be effective in CLIR. For instance, Chinnakotla et. al. [7] studied Multilingual Pseudo-Relevance Feedback (*MultiPRF*): They first translated the query into a target language and then ran retrieval with both source and target language queries to obtain feedback models for both languages. The target language feedback model was then translated back to the source language with a bilingual dictionary, and the resulting model was combined with the original query model and the source feedback model to rank results. Their MultiPRF method was found to be beneficial to IR, and could even improve monolingual retrieval results. The potential usefulness of PRF specifically for patent IR has also been studied [4,5], however solely for monolingual patents.

3 Methodology

The aim of our approach is to turn monolingual queries into multilingual for patent IR. To this end, we extract a translation dictionary of terms and phrases from a parallel patent corpus (Section 3.1). This patent corpus is also the retrieval collection used in this work (see Section 4.1 for its description). We use the extracted dictionary to translate the original monolingual queries (Section 3.2). Section 3.3 describes how we integrate this translation process into retrieval.

3.1 Extracting a Translation Dictionary of Terms and Phrases

Our retrieval collection contains European Patent Office (EPO) patents, which comprise text fields, e.g., *title*, *abstract*, *description*, and *claims*; metadata fields, e.g., *applicant*, *inventor*, *International Patent Classification (IPC)*, and *date published*; and figures and illustrations. The claims field is very important for patent IR, because it contains the legally-binding portion of the patent that may be used later to determine the patent's validity or defend it against infringement [1,2]. In our collection, the claims are manually translated into English, French, and German. Therefore, the claims of our patent collection may be seen as a parallel corpus which can be used to extract translation dictionaries specific to the patent domain.

First, we extract the claims field from documents in our collection that contain claims in all three languages. Then, we align these translated claims. We use a word alignment tool for building our dictionary. However, the sentences in the claims field are often very long [1], and this may cause a problem to the word alignment tool because generally these tools do not handle very long sentences very well; therefore it is necessary to limit the length of sentences. We divide claims into shorter subsentences by splitting sentences using the XML markup found in the patents. This approach is chosen over splitting by non-terminating punctuation (i.e. colon, comma, semicolon, etc.) because punctuation use seems to vary more between languages than the XML markup.

As a result, even though initially the patent claims were perfectly aligned, the subsentences that we get after splitting are not necessarily perfectly aligned, due mainly to variation in the XML markup. To correct this, we align subsentences using the sentence aligner gargantua[2]. We choose gargantua because of its high accuracy in sentence alignment (F_1 scores of 98.47 for aligning German-English sentences and 98.60 for French-English [6]).

---

[2] http://sourceforge.net/projects/gargantua/

Table 1. Patent dictionary: Note that the dictionaries are bidirectional. So there are 1,715,491 French terms that can be translated to an English phrase and 1,715,491 English phrases that can be translated to a French term.

Languages	# entries	$term \rightarrow term$	$term \rightarrow phr.$	$phr. \rightarrow term$	$phr. \rightarrow phr.$
French-English	162,840,175	922,952	1,715,491	3,437,236	156,764,496
French-German	157,977,915	1,645,245	3,570,673	10,419,547	142,342,450
German-English	116,611,676	1,290,111	4,642,276	2,863,824	107,815,465

Having aligned subsentences, we can now start training a statistical machine translation system which will produce a translation dictionary. In this work, we use the state of the art Moses statistical machine translation system [13]. First, we run the GIZA++ word aligner [18] inputting the subsentences we previously aligned with gargantua. GIZA++ produces a word to word alignment, which can also be used as a translation dictionary for terms [11,15,22]. To get the phrase translations, we use the default GDFA (*grow diagonally final AND*) alignment [14] from GIZA++ to train Moses and produce a bidirectional phrase table. The phrase table includes phrase translations with a source phrase of length m and a target phrase of length n, where m and n are between 1-7 inclusive. This means that the phrase table contains term to term translations, term to phrase translations, phrase to term translations, and phrase to phrase translations. The phrase table is also bidirectional, so the size of the German-English dictionary will be the same as the size of the English-German dictionary. The translation probabilities are not symmetric though; e.g., the German word *gefäß* translates to *vessel* with probability 0.61, but *vessel* translates to *gefäß* with probability 0.26.

Table 1 shows the total number of entries for each language pair as well as a breakdown of the total according to type of translation equivalence. Generally, there are many more equivalences involving phrases simply because there are many more phrases than terms in any given language. The highest number of equivalences between a term and a phrase (in either direction) occur for German-English (4,642,276) and French-German (10,419,547) because compounds (which are terms) are frequent in German and they are most often translated as phrases.

3.2   Translating Queries

Given the term and phrase translation dictionary described above, we use two different methods to translate queries; the first is term to term translation (TR_{term}); the second includes phrase translations (TR_{phrase}). In the TR_{term} method, for every term t in a query q, we identify the single best translation t' of the term, and extend q with t'. We define the single best translation to be the most probable one according to the bilingual dictionary extracted by Moses. We do this for all possible source-language combinations, and end up with a multilingual query. An example of term translation can be seen in Table 2, where the first row is the original German query, and the second and third rows are the French and English term translations. This example shows some of the typical problems that occur in automatic term-by-term translations: some terms are

Table 2. Example of term and phrase translation from German to French and English

original German query	ein tintenstrahl aufzeichnungsmaterial mit einem träger und mindestens einer unteren pigment
French term translation	*un jet support avec un support et moins une inférieure pigment*
English term translation	*a inkjet recording with a carrier and least a lower pigment*

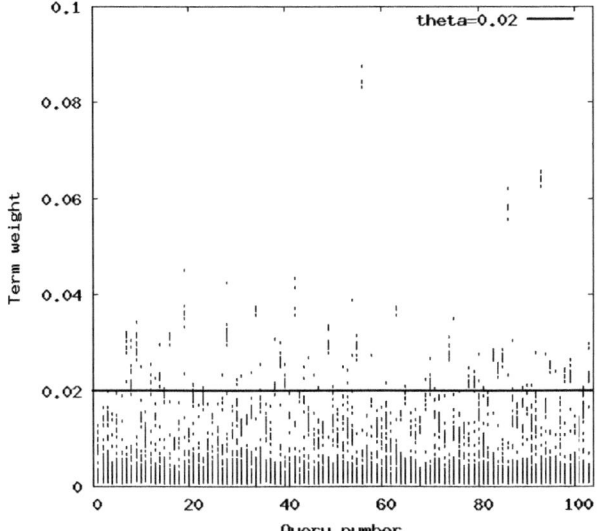

Fig. 1. Term weights of the query set in the baseline run

poorly translated (German "aufzeichnungsmaterial" 'recording material' is translated as "support" by the German-French dictionary) and some terms can only be adequately translated as a phrase (German "aufzeichnungsmaterial" can only be adequately translated as a phrase, "recording material", in English).

In the TR_{phrase} method, we extract only those phrases in the original query for which we have a translation. Our definition of phrase is the longest n-gram (i.e. string of n words) in the dictionary for n=[1-7]. Terms that are not present in any phrase are translated as terms in the multilingual query. Since stopwords will be removed by the retrieval system we do not need to translate phrases of only stopwords. So although "of a" would be considered a phrase in our approach, we do not translate it. Taking this one step further, we remove any stopwords at the beginning or end of a phrase, but preserve the stopwords within phrases. So "of the ink jet" simply becomes "ink jet" while "coated with aluminum" remains the same.

3.3 Translating Salient Terms

In the previous section, we described our query translation method. We do not apply this to all query terms, but to a selection of the most salient terms in the query. The motivation for this is two-fold: on one hand, previous work on query translation for patent IR has showed the limitation of fully translating whole patent queries ([11]); on the other hand, recent work on CLIR has showed that satisfactory retrieval performance can be achieved with approximate translations of queries (e.g. [10]).

We select the terms that are to be translated as follows. Using our baseline retrieval model (see Section 4.1), we look at the term weights assigned to the individual query terms. Figure 1 displays this distribution over our whole query-set (described in Section 4.1). By defining a threshold θ of term weights, we can assume that we reasonably separate the most salient query terms from the rest. Hence, we translate only the terms whose weight $> \theta$. The higher the threshold, the fewer terms are translated. For example, for $\theta = 0.02$, 9.1% of query terms are selected for translation.

4   Experiments

4.1   Experimental Settings

For our experiments we focus on the translation of patent queries from German to English & French and from French to English & German, i.e. we transform an originally monolingual query into a trilingual query. We focus only on French and German as source languages because they have a more complex morphology than English, and hence pose a bigger problem for CLIR than translation from English. We use the patent collection from CLEF-IP 2010 [19] (see Table 3), which has 104 topics in German and French with relevance assessments. These topics are not TREC-style queries, but full patent documents, hence an open problem is how to generate queries out of them [5,23]. We do this by taking the abstract field of the topic patent as the query, following [23]. Our initial query is the set of unique terms from the abstract (average query length= 57.6 terms).

Table 3. CLEF-IP 2010 collection: total size (left) and % by original language (right)

Size	84 GB		French	German
# documents	2,680,604	pct queries	5.0%	29.7%
# tokens	9,840,411,560	pct documents	7.1%	24.0%
# unique terms	20,132,873	pct relevance assessments	5.0%	21.8%

We index the collection using Indri[3] without removing stopwords or stemming. For retrieval we use the Kullback-Leibler language model with Dirichlet smoothing [8]. We tune Dirichlet's μ parameter within $\mu = \{5000, 7500, 10000, 12500, 15000, 17500, 20000\}$. For retrieval, we use standard stoplists for German and French[4]. Our translation approach includes a term weight threshold θ (described in Section 3.3), which we tune: $\theta = \{0.016, 0.02, 0.025\}$. Because our query

---

[3] http://www.lemurproject.org
[4] accessible from http://members.unine.ch/jacques.savoy/clef/

translation approach expands queries with their (partial) translations, we also do runs with pseudo-relevance feedback (PRF), using Indri's default PRF implementation, which is an adaptation of Lavrenko's relevance model [16]. PRF uses these parameters: the number of documents ($fbDocs$) and the number of terms ($fbTerms$). We set $fbDocs = 1$ and $fbTerms = 40$, following [11]. Finally, we include a run that uses Google Translate for query translation. This type of translation differs from ours (we submit the whole query for translation to Google Translate, whereas our approach translates only salient terms/phrases; also, Google Translate is domain-free, whereas our approach uses a patent translation dictionary extracted from the retrieval collection). We include the Google Translate runs simply to contextualize the results from our approach. We use standard TREC evaluation measures: mean average precision (MAP), precision at 10 (P10), and recall. We tune separately for each evaluation measure.

Our experiments are set up as follows:

1. **baseline:** original monolingual query;
2. TR_{term}**:** the original query is expanded with term translations of its most salient terms;
3. TR_{phrase}**:** the original query is expanded with phrase translations of its most salient terms;
4. **PRF:** same as baseline but with PRF;
5. TR_{term}**+PRF:** same as TR_{term} but with PRF;
6. TR_{phrase}**+PRF:** same as TR_{phrase} but with PRF;
7. **Google translate:** the original query is expanded with its full translation using Google translate.

4.2   Experimental Results

Table 4 summarizes our experimental results. For German, term translation seems better than phrase translation. We originally expected that term-to-phrase translations would handle German compounds better than term-to-term translation. Indeed, in query 237, *korngrößenverteilung* translates to *particle size distribution* with phrase translation, and only to *size* using term translation. For this query, term translation improves MAP by 4% over the baseline, and phrase translation improves MAP by 36% over the baseline. However, there are also cases where German is equally well translated with phrases or terms, e.g., in query 201, *tintenstrahl* is translated as *inkjet* or *ink jet*, respectively. For French, phrase translation is clearly better than term translation: MAP and P10 improve substantially over the baseline, compared to term translation. Recall however decreases. For the PRF runs, MAP and P10 are about the same for term and phrase translation, but recall of phrase translation shows a clear improvement compared to term translation. A reason for French benefiting more from phrase translation than German may be that it does not have as many compounds: concepts that are expressed as phrases in French are often translated as compounds in German. E.g., the phrase "flux de matière" 'flux of material' in query 242 gets translated into *materialströme* in phrase translation, and into *material*, *von*, and *strom* in term translation.

Table 4. Best scores in bold. * marks statistical significance p<0.05 using the two-tailed t-test. TR_{term} is term translation and TR_{phrase} is phrase translation

	MAP		P10		Recall	
	German	French	German	French	German	French
baseline	0.0581	0.0527	0.0864	0.0667	0.2456	0.2772
TR_{term}	**0.0598**	0.0556	**0.0875**	0.0867	0.2622	**0.2871**
TR_{phrase}	0.0577	**0.0614**	**0.0875**	*0.1000	0.2671	0.2772
PRF	0.0664	0.0730	**0.0875**	0.1067	0.2661	0.2822
TR_{term}+PRF	0.0667	0.0719	0.0841	0.1000	**0.2749**	0.2871
TR_{phrase}+PRF	**0.0672**	**0.0744**	0.0864	0.1000	0.2739	**0.3218**
Google translate	0.0473	0.0652	0.0659	*0.1200	0.3168	0.3614

On several occasions PRF outperforms the baseline. This is not unexpected, as PRF has been shown to help patent retrieval [4,5]. However, the best result overall for each combination of language and evaluation measure fuses PRF with translation (although in the case of P10 for German, PRF with and without translation are tied). This shows that PRF and translation can contribute different improvements to retrieval performance, and that these two very different approaches are not incompatible. Furthermore, looking at the Google Translate run, it is not surprising that it does best on recall, but underperforms in the other measures: the queries translated by Google contain the full patent abstract and its full translation, meaning that they are very lengthy queries with probably better coverage at the expense of precision.

4.2.1 Analysis by query difficulty. In order to further understand our findings we look more closely at performance on a per-query basis. Specifically, we group queries on the basis of their baseline recall, on the assumption that queries of very low baseline recall will be much more difficult to improve (using either PRF or translation), than queries with higher baseline recall.

Table 5 presents retrieval performance split between three groups of query difficulty: *very hard* (baseline recall = 0%), *hard* (baseline recall = 1%–49%), and *medium* (baseline recall = 50%–100%). We see that German queries of *medium* difficulty underperform with term or phrase translations. On the other hand, results for these runs improve for the *hard*, and *very hard* queries. It seems few queries account for most of this variation. For example, in the group of *medium* queries, one query (query 213) has better recall with term and phrase translation than the baseline, while for two queries (queries 75, 152) the baseline has better recall. The recall for query 152 drops substantially with 8 of 10 relevant documents being retrieved in the baseline and none being retrieved with either word or phrase translation. This single query accounts for most of the decrease in translation results in the *medium* group. Additionally, all of the relevance assessments for this query are German (hence the settings for this query can be seen as biased). A single query can also account for much of the improvement in the *hard* group's translation results. For query 201 for example, 2 of the 20 relevant patent documents are in German and the baseline query only returns those

2 relevant documents. Adding phrase translations (in particular the addition of the phrase "ink jet") increases the number of relevant documents returned to 15. We also observe that *medium* difficulty queries tend to have more relevance judgements in their original source language, and that *hard* queries tend to have relevance judgements from different languages. To the extent that this is the case, it is understandable that results for *medium* queries worsen with translation: a largely monolingual (say, French) result set has high ranks for relevant documents (which are all French), but for a multilingual result set the ranking of some French relevant documents slips. On the other hand, *hard* queries may improve if the baseline monolingual result set (French) does not match many relevant documents (several English documents with a few French), but with the addition of multilingual documents to the result set, more relevant documents are retrieved. This bias of the percentage of a query's relevant documents that are in the original source language of the query also affects the performance of PRF. PRF chooses terms for expansion from the top ranked documents. These documents are likely to be in the same language as the original query. So an original French query in the baseline will expand the query with French terms. In the cases of TR_{term}+PRF and TR_{phrase}+PRF, the highest ranked result with TR is often a multilingual patent document and so the multilingual query will have multilingual expansions. With this multilingual patent collection, multilingual query expansion should be more desirable, and in fact TR+PRF outperforms PRF for MAP and recall. In particular, TR+PRF does better than PRF for *hard* and *very hard* queries where it appears there is a larger percentage of relevant documents in a language different than the query.

In contrast to the German results, the French MAP and P10 scores improve for term and phrase translation across the *hard* and *medium* groups, and results for *very hard* remain the same. Query 239 is an example of a query which improves for MAP and P10 while recall remains the same. The original French query has recall of 1, returning all four of its relevant documents (two of them include translated claims). Phrase translation still proves to be useful here in improving the relevant documents' ranking (MAP and P10 both improve). For recall, we see the same behavior as with German: recall drops using translations (TR_{term} and TR_{phrase}) in the *medium* group and rises for harder queries. For the *medium* group, only one less document is retrieved using TR_{term} with respect to the baseline. This single document accounts for the 3.6% drop in recall. Note that there are only 15 French queries and an average of 13.5 relevance assessments per query, so one relevant document being added to or dropped from the result set can have a big impact. Table 6 shows that for the majority of queries, recall remains the same.

Furthermore, Table 6 shows how TR_{term} and TR_{phrase} performed against the baseline for individual queries. We counted, for each evaluation measure, the number of queries that performed better than, worse than, or equal to the baseline. All measures for both translation methods, with the exception of recall for TR_{term}, have more queries that exceed the baseline than queries that drop below it. We observe that, for each language/evaluation measure combination, there

Table 5. German and French results by difficulty. * marks statistical significance p<0.05 using the two-tailed t-test.

	German				French			
	hard++ (31.8%)	hard (48.9%)	medium (19.3%)	all (100%)	hard++ (20.0%)	hard (46.7%)	medium (33.3%)	all (100%)
	MAP							
baseline	0.0000	0.0375	**0.2058**	0.0581	0.0000	0.0676	0.0635	0.0527
TR_{term}	0.0002	**0.0417**	0.2036	**0.0598**	0.0000	0.0694	0.0695	0.0556
TR_{phrase}	**0.0003**	0.0406	0.1953	0.0577	0.0000	**0.0794**	**0.0729**	**0.0614**
PRF baseline	0.0013	0.0525	**0.2091**	0.0664	0.0000	**0.0907**	0.0919	0.0730
TR_{term}+PRF	**0.0018**	0.0543	0.2049	0.0667	0.0000	0.0894	0.0905	0.0719
TR_{phrase}+PRF	0.0014	**0.0558**	0.2045	**0.0672**	0.0000	0.0852	**0.1038**	**0.0744**
	P10							
baseline	0.0000	0.0721	**0.2647**	0.0864	0.0000	**0.0857**	0.0800	0.0667
TR_{term}	0.0000	0.0767	0.2588	**0.0875**	0.0000	**0.1286**	0.0800	0.0867
TR_{phrase}	0.0000	**0.0837**	0.2412	**0.0875**	0.0000	**0.1286**	0.1200	*0.1000
PRF baseline	0.0000	0.0860	**0.2353**	0.0875	0.0000	0.1143	0.1600	0.1067
TR_{term}+PRF	0.0000	0.0837	0.2235	0.0841	0.0000	0.1143	0.1400	0.1000
TR_{phrase}+PRF	0.0000	**0.0907**	0.2176	0.0864	0.0000	0.1000	**0.1600**	0.1000
	Recall							
baseline	0.0000	0.2532	**0.6328**	0.2456	0.0000	0.2301	**0.5882**	0.2772
TR_{term}	**0.0405**	0.2731	0.5989	0.2622	0.0000	**0.2566**	0.5686	**0.2871**
TR_{phrase}	0.0372	**0.2821**	0.6045	**0.2671**	0.0263	0.2389	0.5490	0.2772
PRF baseline	0.0372	0.2712	**0.6328**	0.2661	0.0263	0.2301	0.5882	0.2822
TR_{term}+PRF	**0.0642**	**0.2857**	0.5932	**0.2749**	0.0263	0.2301	**0.6078**	0.2871
TR_{phrase}+PRF	0.0608	0.2821	0.6045	0.2739	**0.0526**	**0.2920**	0.5882	**0.3218**

are more queries improved by phrase translation than queries improved by term translation (with one tie, 2-2, for French P10). However, in three cases (German MAP, German recall, French recall), there are also more queries where phrase translation does worse than term translation; the other three cases (German P10, French MAP, French P10) are tied. Our interpretation of these results is that phrases have significant potential for improving retrieval results, but they have to be carefully selected, otherwise performance will deteriorate. In contrast, term translations are more conservative and less likely to have a negative effect, but at the same time they offer limited improvements.

4.2.2 Tuning of Dirichlet parameter μ. Finally, Figures 2-3 show MAP and P10 scores across the tuning range of μ. The more stable the line of our approach, the less sensitive it is to factors pertaining to variation in document length and collection statistics. For the MAP tuning for both German and French, the results for term and phrase translation are quite stable, while the three runs that use PRF drop (between 10000 and 12500 for German and between 12500 and 15000 for French).

Table 6. German and French performance per query. $TR_{\text{term}} > baseline$ indicates that the evaluation measure was greater for term translation than the baseline. $<$ and $=$ indicate less than and equal to, respectively. Other notation as in Table 4.

Eval. meas.	$TR_{\text{term}} >$ baseline	$TR_{\text{term}} <$ baseline	$TR_{\text{term}} =$ baseline	$TR_{\text{phrase}} >$ baseline	$TR_{\text{phrase}} <$ baseline	$TR_{\text{phrase}} =$ baseline
German						
MAP	19	13	56	33	23	32
P10	5	4	79	7	4	77
recall	16	9	63	18	12	58
French						
MAP	5	3	7	9	3	3
P10	2	0	13	2	0	13
recall	2	2	11	4	3	8

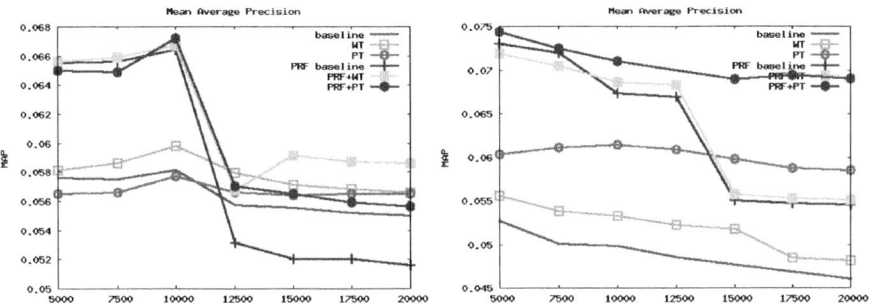

Fig. 2. Dirichlet prior μ tuning for German (left) and French (right) versus MAP (y axis)

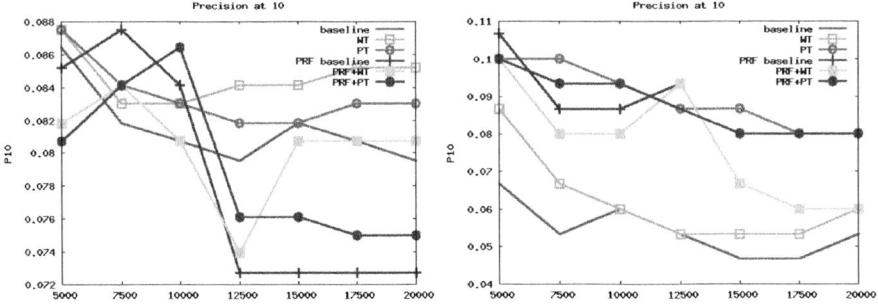

Fig. 3. Dirichlet prior μ tuning for German (left) and French (right) versus P10 (y axis)

Note that results are less stable for the P10 tuning, although word and phrase translations appear more stable than PRF runs. The German PRF results drop for P10 like they did for MAP.

5 Conclusions

In an increasingly networked world, problems of multilinguality are gaining importance in information retrieval (IR). Most IR approaches in multilingual settings use some form of translation. In this paper, we adopted an expansion approach to translation for patent IR, where translations of query parts are added as additional terms to the query. We looked at two alternative translation methods, term translation and phrase translation. Our experimental evaluation showed good results for both, especially on hard queries. Phrase translation seems to be more beneficial for French than for German because German often uses single-term compounds instead of phrases, thus limiting the potential benefit of phrase to term and phrase to phrase translations.

References

1. Atkinson, K.H.: Toward a more rational patent search paradigm. In: 1st ACM Workshop on Patent IR, pp. 37–40 (2008)
2. Azzopardi, L., Vanderbauwhede, W., Joho, H.: Search system requirements of patent analysts. In: SIGIR, pp. 775–776 (2010)
3. Ballesteros, L., Croft, W.B.: Phrasal translation and query expansion techniques for cross-language information retrieval. In: SIGIR, pp. 84–91 (1997)
4. Bashir, S., Rauber, A.: Improving retrievability of patents with cluster-based pseudo-relevance feedback documents selection. In: CIKM, pp. 1863–1866 (2009)
5. Bashir, S., Rauber, A.: Improving retrievability of patents in prior-art search. In: Gurrin, C., He, Y., Kazai, G., Kruschwitz, U., Little, S., Roelleke, T., Rüger, S., van Rijsbergen, K. (eds.) ECIR 2010. LNCS, vol. 5993, pp. 457–470. Springer, Heidelberg (2010)
6. Braune, F., Fraser, A.: Improved unsupervised sentence alignment for symmetrical and asymmetrical parallel corpora. In: COLING (2010)
7. Chinnakotla, M.K., Raman, K., Bhattacharyya, P.: Multilingual prf: english lends a helping hand. In: SIGIR, pp. 659–666 (2010)
8. Croft, W.B., Lafferty, J.: Language Modeling for Information Retrieval. Kluwer Academic Publishers, Dordrecht (2003)
9. Fujii, A., Utiyama, M., Yamamoto, M., Utsuro, T.: Overview of the patent translation task at the NTCIR-7 workshop. In: NTCIR (2008)
10. Gao, W., Niu, C., Nie, J.-Y., Zhou, M., Wong, K.-F., Hon, H.-W.: Exploiting query logs for cross-lingual query suggestions. TOIS 28(2) (2010)
11. Jochim, C., Lioma, C., Schütze, H., Koch, S., Ertl, T.: Preliminary study into query translation for patent retrieval. In: PaIR, Toronto, Canada. ACM, New York (2010)
12. Kettunen, K.: Choosing the best MT programs for CLIR purposes – can MT metrics be helpful? In: Boughanem, M., Berrut, C., Mothe, J., Soule-Dupuy, C. (eds.) ECIR 2009. LNCS, vol. 5478, pp. 706–712. Springer, Heidelberg (2009)
13. Koehn, P., Hoang, H., Birch, A., Callison-Burch, C., Federico, M., Bertoldi, N., Cowan, B., Shen, W., Moran, C., Zens, R., Dyer, C., Bojar, O., Constantin, A., Herbst, E.: Moses: open source toolkit for statistical machine translation. In: ACL, pp. 177–180 (2007)
14. Koehn, P., Och, F.J., Marcu, D.: Statistical phrase-based translation. In: NAACL, pp. 48–54 (2003)

15. Larkey, L.S., Connell, M.E.: Structured queries, language modeling, and relevance modeling in cross-language information retrieval. Inf. Process. Manage. 41(3), 457–473 (2005), doi:10.1016/j.ipm.2004.06.008
16. Lavrenko, V., Croft, W.B.: Relevance-based language models. In: SIGIR, pp. 120–127 (2001)
17. Oard, D.W., Diekema, A.R.: Cross-language information retrieval. Annual Review of Information Science and Technology 33, 223–256 (1998)
18. Och, F.J., Ney, H.: A systematic comparison of various statistical alignment models. Computational Linguistics 29(1), 19–51 (2003)
19. Roda, G., Tait, J., Piroi, F., Zenz, V.: CLEF-IP 2009: Retrieval experiments in the intellectual property domain. In: Peters, C., Di Nunzio, G.M., Kurimo, M., Mostefa, D., Penas, A., Roda, G. (eds.) CLEF 2009. LNCS, vol. 6241, pp. 385–409. Springer, Heidelberg (2010)
20. Tait, J. (ed.): 1st ACM Workshop on Patent IR (2008)
21. Tait, J. (ed.): 2nd ACM Workshop on Patent IR (2009)
22. Wang, J., Oard, D.W.: Combining bidirectional translation and synonymy for cross-language information retrieval. In: SIGIR, pp. 202–209 (2006)
23. Xue, X., Croft, W.B.: Automatic query generation for patent search. In: CIKM, pp. 2037–2040 (2009)

Supporting Arabic Cross-Lingual Retrieval Using Contextual Information

Farag Ahmed, Andreas Nürnberger, and Marcus Nitsche

Data & Knowledge Engineering Group
Faculty of Computer Science
Otto-von-Guericke-University of Magdeburg
{farag.ahmed,andreas.nuernberger,marcus.nitsche}@ovgu.de
http://www.findke.ovgu.de/

Abstract. One of the main problems that impact the performance of cross-language information retrieval (CLIR) systems is how to disambiguate translations and - since this usually can not be done completely automatic - how to smoothly integrate a user in this disambiguation process. In order to ensure that a user has a certain confidence in selecting a translation she/he possibly can not even read or understand, we have to make sure that the system has provided sufficient information about translation alternatives and their meaning. In this paper, we present a CLIR tool that automatically translates the user query and provides possibilities to interactively select relevant terms using contextual information. This information is obtained from a parallel corpus to describe the translation in the user's query language. Furthermore, a user study was conducted to identify weaknesses in both disambiguation algorithm and interface design. The outcome of this user study leads to a much clearer view of how and what CLIR should offer to users.

Keywords: cross lingual information retrieval, word sense disambiguation.

1 Introduction

In a time of wide availability of communication technologies, language barriers are a serious issue to world communication and to economic or cultural exchange. More comprehensive tools to overcome such barriers, such as machine translation and cross-lingual information retrieval applications, are currently in strong demand. The increasing diversity of web sites has created millions of multilingual resources in the World Wide Web. At a first glance, it seems that more information can be retrieved by non-English speaking people. However, even users that are fluent in different languages face difficulties when they try to retrieve documents that are not written in their mother tongue or if they would like to search documents in all languages they can speak in order to cover more resources with a single query. The design of a multilingual information system faces specific challenges regarding the best way of handling multiple languages, best query

A. Hanbury, A. Rauber, and A.P. de Vries (Eds.): IRFC 2011, LNCS 6653, pp. 30–45, 2011.
© Springer-Verlag Berlin Heidelberg 2011

translations and ergonomic requirements for the user. In the past, most research has been focused on the retrieval effectiveness of CLIR through Information Retrieval (IR) test collection approaches [7], whereas few researchers have been focused on giving the user a confidence in the translation that the user can not read or understand [16]. Cross-lingual information retrieval (CLIR) adds a way to efficiently get information across languages. However, to achieve this goal, the limitations imposed by the language barriers are a serious issue e.g. translation disambiguation, word inflection, compound words, proper names, spelling variants and special terms [2]. In order to get information across languages, the user query needs to be translated. This translation is not a trivial task, especially for some language such as Arabic. Arabic poses a real translation challenge for many reasons e.g. translation cannot be performed without a pre-processing step. Arabic is different from English and other Indo-European languages with respect to a number of important aspects: words are written from right to left; it is mainly a consonantal language in its written forms, i.e. it excludes vowels; its two main parts of speech are the verb and the noun in that word order, and these consist, for the main part, of triliteral roots (three consonants forming the basis of noun forms that are derived from them). Arabic is a morphologically complex language, in that it provides flexibility in word formation (inflection) making it possible to form hundreds of words from one root. Another issue for the Arabic language is the absence of diacritics (sometimes called voweling). Diacritics can be defined as symbols over and under letters, which are used to indicate the proper pronunciations, hence also define the meaning of a word and therefore have important disambiguating properties. The absence of diacritics in Arabic texts poses a real challenge for Arabic natural language processing as well as for translation, leading to high ambiguity. Though the use of diacritics is extremely important for readability and understanding, diacritics is very rarely used in real life situations. Diacritics don't appear in most printed media in Arabic regions nor on Arabic internet web sites. They are visible in religious texts such as the Quran, which is fully diacritized in order to prevent misinterpretation. Furthermore, the diacritics are present in children's books in school for learning purposes. For native speakers, the absence of diacritics is not an issue. They can easily understand the exact meaning of the word from the context, but for inexperienced learners as well as in computer usage, the absence of the diacritics is a real issue. When the texts are unvocalized, it is possible that several words have the same form but different meaning. For more overview about Arabic language issues and approaches to tackle them, we refer the reader to our previous works [3] and [4]. In the following we give brief overview about the state-of the art CLIR interaction tools clarifying some of their limitations.

2   Related Work in CLIR Tools

The issues of CLIR have been discussed for several decades. In the early seventies experiments for retrieving information across languages were first initiated by Salton [18]. CLIR systems allow the user to submit the query in one language and retrieve the results in different languages and thus provide an important capability that can help users to meet this challenge. In addition to the classical IR tasks, CLIR also requires that the query (or the documents) be translated from one language into another. Query translation is the most widely used technique for CLIR due to it is low computationally cost for translation compared to the effort of translating a large set of documents. However, the query translation approach suffers from translation ambiguity as queries are often short and do not provide rich context for disambiguation. CLIR systems are typically divided into two main approaches: First, systems that exploit explicit representations of translation knowledge such as bilingual dictionaries or machine translation (MT), e.g., [6,11] and second, systems that automatically extract useful translation knowledge from comparable or parallel corpora using statistical/probabilistic models, e.g., [6]. For example, in [19] Brown proposed an approach to construct a thesaurus based on translating the word in the original query then counting its co-occurrences information and storing it with the corresponding word in the target language.

For a dictionary based approach, one can use a general-purpose dictionary or a special dictionary for a specific task, e.g., a medical terminology dictionary for translation. A clear drawback to this approach is that one word might have multiple translations (meanings) in the target language and thus it is very difficult to determine the correct meaning that should be choosen for the translation. Furthermore, a dictionary does not contain all words, e.g., technical terms or proper names. Using a statistical/probabilistic model, based on parallel text, a dictionary translation can be automatically improved because related cross lingual word pairs appear in similar context in such a collection. However, corpora-based approaches for CLIR suffer usually from a domain specific drawback due to the limited coverage of the used corpora. In order to overcome such limitation, hand-crafted lexical resources can lead to significant performance improvement. Lexical resources are irreplaceable for every natural language processing (NLP) system. For example, in order to improve the performance of word sense disambiguation applications, an adequate lexical resource is necessary. In [8] Clough and Stevenson dealt with the problem of selecting the correct translation out of several translations obtained from the bilingual dictionary by the use of EuroWordNet [20].

Despite the clear effort which has been directed toward retrieval functionality and effectiveness, only little attention was paid to developing multilingual interaction tools, where users *are really considered* as an integral part of the retrieval process. One potential interpretation of this problem is that users of CLIR might not have sufficient knowledge of the target languages and therefore they are usually not involved in multilingual processes [17]. However, the involvement of the user in CLIR systems, by reviewing and amending the query, have been studied, e.g. the *Keizai* system [16], a Multilingual Information Retrieval Tool *UCLIR* [1], the Maryland Interactive Retrieval Advanced Cross-Language Engine *MIRACLE* system [15], and the MultiLexExplorer [12]. In the following we describe these CLIR tools in detail.

In the Keizai project [16], the goal was to design a Web-based cross-language text retrieval system that accepts the query in English and searches Japanese and

Korean web data. Furthermore, the system displays English summaries of the top ranked retrieved documents. The query terms are translated into Japanese or Korean languages along with their English definitions and thus this feature allows the user to disambiguate the translations. Based on the English definitions of the translated query terms, the user who does not understand the Japanese or Korean language can select the appropriate one, out of several possible translations. Once the user selects those translations whose definitions are consistent with the information needed, the search can be performed. Keizai has some shortcomings, for example, in order to select the appropriate translation (translation disambiguation) the user has to check all possible translation definitions before the user can select the appropriate translation. This can be very laborious especially for query terms that have abundant possible translations. In addition to this, Keizai relies on the use of a bilingual dictionary for translation as well as for disambiguation. However, bilingual dictionaries in which the definitions of source language (English in this case) are available for each translation for the target languages are very rare.

The UCLIR system [1] performs its task in any of the following three different modes: the first mode, using a multilingual query (query can consist of terms of different languages), the second mode using an English query without user involvement in the multilingual query formulation, the third mode using an English query with user involvement in the formulation of the multilingual queries. In UCLIR the user has to check all translation alternatives and select the appropriate one, which can be laborious. Furthermore, wide coverage dictionaries are very essential in achieving any significant success.

In MIRACLE [15], in order to support the interactive CLIR, the system uses the *user-assisted query translation*. The user-assisted query translation gives the user an opportunity to be involved in the translation process by interacting with the tool to select relevant terms. The user assisted-query translation feature supports the user to select or delete a translation from the suggested one. In case the user deleted a corrected translation, the system reacts, in that the searcher can see the effect of the choice and have possibilities to learn better control of the system. This is done by providing the following features, the meaning of the translation (loan word or proper name), using back translation, a list of possible synonyms are provided. Translation examples of usage are obtained from translated or topically-related text. In MIRACLE, there are two types of query translations, fully automatic query translation (using machine translation) and user-assisted query translation. In fully automatic translation the user can be involved only once. After the system translates the query and retrieves the search results, the user can refine the query if the user is not satisfied with the search results. The rapid adaption to new languages was taken into account in the design of the MIRACLE system. The query language is always English, in MIRACLE. However, language resources that are available for English can be leveraged, regardless of the document language. Currently, MIRACLE works with a simple bilingual term list. However, it is designed to readily leverage additional resources when they are available. Although MIRACLE overcomes some of the limitations of the previously mentioned CLIR interaction tools, it also has some limitations. For example, despite the use of automatic translation in MIRACLE, the user has no influence on refining the translation before the search can be conducted e.g. providing contextual information that describes the translation in the user's own language, in that the user can have a certain degree of confidence in the translation. In addition, to the previously mentioned limitation, in MIRACLE, single word translations are used, which forces the user to spend a lot of effort checking each single translation alternative with their meanings before the user can select/deselect translations.

In MultiLexExplorer [12], the goal was to support multilingual users in performing their web search. Furthermore, the MultiLexExplorer supports the user in disambiguating word meanings by providing the user with information about the distribution of words in the web. The tool allows users to explore combinations of query term translations by visualizing EuroWordNet [1] relations together with search results and search statistics obtained from web search engines. Based on the EuroWordNet, the tool supports the user with the following functionality:

- exploring the context of a given word in the general hierarchy,
- searching in different languages, e.g. by translating word senses using the interlingual index of EuroWordNet,
- disambiguating word sense for combinations of words,
- provide the user with the possibility to interact with the system, e.g. changing the search word and the number of retrieved documents,
- expanding the original query with extra relevant terms, and in
- automatically categorizing the retrieved web documents.

Despite the advanced visual and functional design of MultiLexExplorer, it relies on the use of EuroWordNet, which only employs a limited number of languages. Furthermore, no automatic translation is integrated into the tool instead the user has to check many word sense combinations.

3   The Proposed Interface

The proposed interface provides users with interactive contextual information that describes the translation in the user's query language, in order to allow users to have a certain degree of confidence in the translation. In order to consider users as an integral part of the retrieval process, the tool provides the users with possibilities to interact, i.e. they can select relevant terms from the contextual information in order to improve the translation and thus improve the CLIR process. The proposed tool deals with two issues: Firstly, there is translation ambiguity, where one word in one language can have several meanings in another language, and secondly, the user's insufficient linguistic proficiency of the target language. To tackle the former, the proposed interface makes use of automatic query translation that includes a statistical disambiguation process (in the above mentioned tools, the disambiguation was done by the user). To tackle the latter

---

[1] http://www.illc.uva.nl/EuroWordNet/

issue, the tool supports the user by providing her/him with interactive contextual information that describes the translation in the query language. Based on the provided interaction methods, the user can iteratively improve the translation and thus improve the overall CLIR process. In the following, we describe in more detail how we dealt with the translation ambiguity and with the user's lack of proficiency in the target language.

3.1   Automatic Translation

The automatic translation method consists of two main steps: First, using an Arabic analyzer, the query terms are analyzed and the senses (possible translations) of the ambiguous query terms are identified. Second, the most likely correct senses of the ambiguous query terms are selected based on co-occurrence statistics. We used cohesion scores for possible translation-candidate pairs (translation combinations) to resolve the translation ambiguity of the user's query terms. However, this approach is affected by the sparseness of translation combinations in the underlying corpus. One poorly distributed term can affect the whole cohesion scores obtained from the corpus and therefore in some cases only few - and thus unreliable - statistical co-occurrence data is available or in the worst case none at all. In order to obtain robust disambiguation methods, this data sparseness issue needs to be tackled (see Section 3.1.2).

3.1.1   Mutual Information (MI)

Giving a source of data, Mutual Information (MI) is a measure to calculate the correlation between terms in specific space (corpus or web). MI based approaches have been used often in word sense disambiguation task e.g. [10]. The automatic translation process starts by translating each query term independently. This is done by obtaining a set of possible translations of each of the query terms from the dictionary. Based on the translation sets of each term, sets of all possible combinations between terms in the translation sets are generated. Using co-occurrence data extracted from monolingual corpus (English Gigaword Corpus) [2] or Web, the translations are then ranked based on a cohesion score computed using Mutual Information: Given a query $q = \{q_1, q_1, ..., q_n\}$, and its translation set $S_{qk} = \{q_k, t_i\}$, where $1 \leq k \leq n, 1 \leq i \leq m_k$ and m_k is the number of translations for query term k. The MI score of each translation combination can be computed as follows:

$$MI(q_{t_1}, q_{t_2}, ..., q_{t_n}) = log \frac{P(q_{t_1}, q_{t_2}, ..., q_{t_n})}{p(q_{t_1})p(q_{t_2})...p(q_{t_n})} \quad (1)$$

with $P(q_{t_1}, q_{t_2}, ..., q_{t_n})$ being the joint probability of all translated query terms to occur together, which is estimated by counting how many times $q_{t_1}, q_{t_2}, ..., q_{t_n}$ occur together in the corpus. The probabilities $p(q_{t_1})p(q_{t_2})...p(q_{t_n})$ are estimated by counting the number of individual occurrences of each possible translated query term in the corpus.

---

[2] www.ldc.upenn.edu/

A Walkthrough Example. To clarify how the approach works, let's consider the following query (العالية الصحة منظمة mnẓmh ālṣhh ālʿālmīh , "World Health Organization"). The algorithm retrieves a set of possible translation for each query term $S_{qk} = \{q_k, t_i\}$ for each query term q_m from a dictionary [3]. The term منظمة mnẓmh has the following transaltions ("organization", "organized", "orderly", "arranged", "organizer", "sponsor"), the query term الصحة ālṣhh has the following translation ("health", "truth", "correctness") and the query term العالمية ālʿālmīh has the following translation ("universality", "internationalism", "international", "world", "wide"). For example, we are considering the first query term (منظمة mnẓmh , "organization"), that has six translations ("organization", "organized", "orderly", "arranged", "organizer", "sponsor"). The set of translations is thus defined with $k = 1$ and $1 \leq i \leq 6$ as $S_{q1} = \{q_{1,t_1}, q_{1,t_2}, ..., q_{1,t_6}\}$. The translation sets for all query terms are retrieved from the bilingual dictionary. After the translation sets are retrieved, the translation combinations between the translations for each of the query terms are created. The total number of combinations can be computed by simply multiplying the sizes of all translation sets. For the previous example we thus obtain a total number of combinations $6 \times 3 \times 5 = 90$ as listed in Table 1 (only the first ten combinations are listed here). Finally, the MI score will be calculated for all possible combinations of the translation-candidate pairs (translation sets). The translation combination that maximizes the MI score will be selected as the best translation for the user query. Before we present an evaluation of this approach, we first discuss one of its main drawbacks, the data sparseness issue and how we dealt with it, in the following.

Table 1. Terms and their average MI Scores

S/N	Translation Combinations	Occurrence	MI Score
1	organization AND health AND world	5579	8,62651
2	organization AND health AND international	2457	7,80648
3	organized AND health AND world	415	6,0282
4	organized AND health AND international	328	5,79295
5	organization AND truth AND world	229	5,43367
6	organization AND health AND wide	225	5,41608
7	organization AND truth AND international	205	5,32297
8	arranged AND health AND world	137	4,91995
9	sponsor AND health AND world	116	4,75357
10	organized AND truth AND world	99	4,59511
...	...	...	...

3.1.2   Revised Mutual Information (MI)

In order to clarify the data sparseness issue, let's consider the following example. When translating the Arabic query "الادوية مبيعات ضريبة ḍrībh mbīāt ālādwīh " ("medications tax sales"), there might be not enough statistical co-occurrences

---

[3] http://www.nongnu.org/aramorph/english/index.html

Table 2. Translation combination for " ضريبة مبيعات الادوية‏ *ḍrībh mbīʿāt āladwīh* "

S/N	Translation Combinations
1	tax AND sold AND remedies
2	tax AND sold AND medications
3	tax AND sales AND remedies
4	tax AND sales AND medications
5	levy AND sold AND remedies
6	levy AND sold AND medications
7	levy AND sales AND remedies
8	levy AND sales AND medications

data obtained from the corpus and thus the algorithm will fail to translate this query. However, the revised algorithm can exploit the corpus and check out which term has no cohesion score with other terms, i.e. terms that do not cooccure with others and thus this term can be detected and eliminated. In this case, the term that affects the cohesion score is " الادوية‏ *āladwīh* " ("the medications", "the remedies") and eliminating this term will allow to obtain sufficient statistical co-occurrence data. The rest of the terms are " ضريبة مبيعات‏ *ḍrībh mbīʿāt* " ("tax sales") have very high cohesion score due to the fact that these terms are widely available in the corpus. The noise term detection process will be performed only if the proposed disambiguation algorithm failed to provide a translation due to the lack of statistical co-occurrence data for the query terms as a whole. In the following, we describe how the elimination process is performed by the algorithm.

Step 1. The algorithm generates all possible translation combinations: Given the user query $Q = \{t_1, t_2, ..., t_n\}$ " ضريبة مبيعات الادوية‏ *ḍrībh mbīʿāt āladwīh* " ("medications tax sales"), the set of possible translation combinations is $\{Tcom_1, Tcom_2, ..., Tcom_n\}$, where n defines the number of possible translation combinations for the user query Q. In our example, $n = 8$, so 8 translation combinations are generated (See Table 2).

Step 2. The algorithm constructs possible term combinations between the generated translation combinations: Given a translation combination $Tcom_i = \{t_1, t_2, ..., t_n\}$, we compute its possible term combinations as follows: Given the set of $\frac{n(n-1)}{S}$ combinations, where S is the size of each combination set, n is the number of terms in the given translation combination. The set of term combinations between all translation combination terms is defined as $Com_i = \{\{Tcom_{i,j}, Tcom_{i,k}\} | 1 \le j < n, j < k \le n\}$. Let's consider the translation combination ("tax" AND "sales" AND "medications") number 4. Here $i = 4$, $n = 3$ and $S = 2$. After generating all possible combinations between the translation combination terms (See Table 3), the mutual information score for each term combination will be calculated based on Eq 1.

Step 3. The algorithm computes the MI score for each individual term combination, and then the MI score mean will be calculated. The term that has the lowest MI score, which is below the MI core mean, will be considered as a noise term and thus the term combination that includes this term will be eliminated. As shown in Table 3, term combinations with the term ("medications") always have the lowest MI score (1.38629 and 3.98898).

Table 3. Term Combinations and their MI Scores

S/N	Term combinations	MI Score
1	tax AND sales	8.86319
2	tax AND medications	1.38629
3	sales AND medications	3.98898
	The MI score mean	4.746

Step 4. The algorithm calculates the average MI score individually for all terms in the constructed term combinations and compares them with the MI score mean. As shown in Table 4, the term ("medications") has the lowest MI average score (2.687), which is below the MI score mean (4.746), and thus will be classified as a noise term and will be eliminated. In contrast, all other terms have an average mutual score, which is above the MI score mean and thus have significant statistical co-occurrence data needed for translation.

Table 4. Terms and their average MI Scores

S/N	Term	MI average Score
1	sales	6.426
2	tax	5.124
3	medications	2.687

Step 5. Using the dictionary, possible translations with their contextual information for the noise term ("medications") will be suggested. Based on the fact that some dictionaries have a ranked translation (usually the translations provided by dictionaries are presented in an ordered way, based on their common use and thus the most common translation is listed first) [5], the algorithm will display the suggested translations in ranked order. Ultimately, if the user agrees with the translation of the noise term based on the contextual information, the translated noise term will be included in the translation, otherwise the translated noise term will be cancelled.

3.2   Interactive Contextual Information

In order to help the user to better understand the meaning of the different query term translations, the tool provides contextual information to clarify the usage - and thus the meaning - of the terms. Figure 1 (a) shows an example, where the user submits the Arabic query " الحكومة دين‎ dyen alhkwmh ". The query is automatically translated and the best three translation will be displayed to the user in ranked order (see Fig. 1 (b)). Each translation is looked up in the target language documents index (one translation after the other) in order to obtain the relevant documents (contextual information), for the translation. In order to get the equivalent documents in the source language, the parallel corpus (taken from www.ldc.upenn.edu/) is queried. Since it is possible that some retrieved documents will be very similar – which would result in duplicate contextual information – the documents retrieved from the source language are automatically grouped and contextual information is selected only once from each cluster. The final selected contextual information is not provided to the user as raw text as it is the case in the state-of-the art CLIR tools, but instead, it will be presented as a classified representation of each contextual information term: each term of the contextual information is color-coded according to its related type and can be selected as a disambiguating term (see Fig. 1 (c)).

In order to clarify the interaction scenario, we consider the submitted user query " الحكومة دين‎ dyen alhkwmh ". The query term " الحكومة‎ alhkwmh " has two translations ("the government" or "the administration"), while the other term " دين‎ dyen " has several possible translations e.g. ("religion" or "debt"). Based on the MI score, translation alternatives are displayed in ranked order together with their contextual information (see Fig. 1 (b) and (c)). Thus the user has the possibility to select the suitable translation. Here, the translations provided

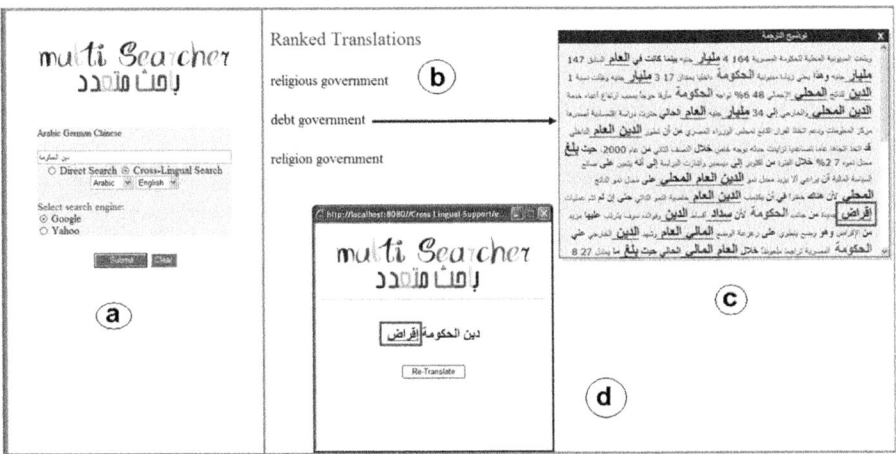

Fig. 1. The translation alternatives with their contexual information. In the contextual information window (c) the user's query terms are displayed in green, suggested terms, by the tool based on highly frequent co-occurrences in the context of the query are in bold blue and underlined, all remaining terms are blue except stop words that are black and not selectable.

by the system ("the government religion") and ("the government debt") are correct even though they are used in a different context. This is due to the fact that ("government") appears frequently in the context of ("religion" or "debt"). As shown in Fig. 1 (b) and (c), the user is interested in the second ranked translation ("debt government"). Using the contextual information, the user can select one or more terms to improve the translation. To simplify the user's task, the tool automatically proposed relevant terms (highlighted in bold blue and underlined), e.g. ("payment", "financial", "lending", "loan"). Once the user selects, for example, the interactive term "اقراض $\bar{a}qr\bar{a}\dot{d}$ " ("lending"), the tool re-translates the modified query and displays the new translations ("debt government loan", "debt government lending" and "debt administration loan"), to the user (see Fig. 1 (d)). The user can, with a simple mouse click, confirm the translation which will then be sent to his favorite search engine using integrated web services, e.g. Yahoo or Google, retrieving the results and displaying them (see Fig. 2 for yahoo example).

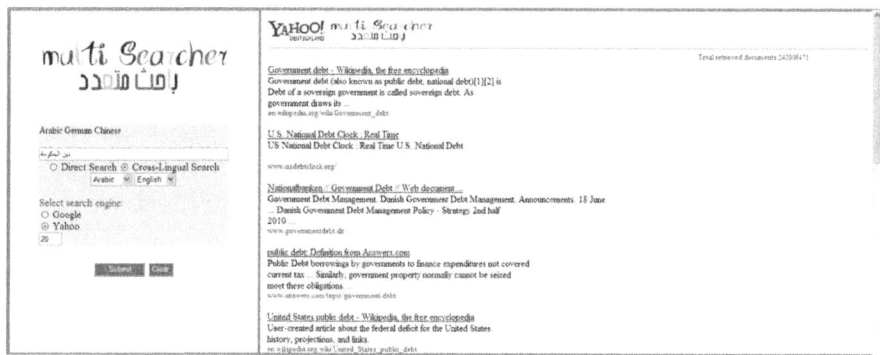

Fig. 2. Relevant documents for the desired translation obtained from web (Yahoo)

4   Evaluation

Different from the mentioned state-of-the art tools where the evaluation was done based on the relevancy of the documents retrieved from local collection, i.e. using recall and precision, we are focusing on evaluating the automatic translation algorithm integrated in the tool. The retrieval process itself is performed by standard search engine web services called by the tool after the translation was performed. In other words, the main goal of our tool is to support the user in performing the CLIR task by providing him/her with contextual information that describe the translation in the user's own language. Therefore, we had to use different measures that are more appropriate to evaluate this specific task. Furthermore, we evaluated the tool from the user point of view in the form of user study.

Table 5. Tool overall performance using monolingual corpus and the web

Statistical data source	Applicability	Precision
Monolingual corpus	75%	70%
WEB	90%	85%

4.1   Tool Accuracy: Using Corpus and Web

We evaluated the proposed method, by conducting two experiments. In the first experiment co-occurrence data was used, which was obtained from the monolingual corpus and the other was based on the co-occurrence data, which was obtained from the web using a particular search engine (here, Yahoo). The English Gigaword corpus [4] is a comprehensive archive of newswire text data that has been acquired over several years by the Linguistic Data Consortium (LDC) at the University of Pennsylvania. We used the third edition of the English Gigaword Corpus. In order to evaluate our proposed tool, 20 Arabic queries were randomly selected for the test users from the corpus. These queries included at least one ambiguous word which has multiple English translations. The number of senses per test word ranged from 1 to 17, and the average was 4.9. The number of query translation combinations ranged from 4 to 215 with the average being 31.3. In order to evaluate the performance of the disambiguation algorithm, we used two measurements: applicability and precision [9]. The translation can be considered as correct if it is one of the three suggested translation. The applicability is the proportion of the ambiguous words that the algorithm could disambiguate. The precision is the proportion of the correct disambiguated senses for the ambiguous word. Table 5 shows, the applicability and precision of the proposed algorithm, using monolingual corpus, over the 20 test queries. The applicability and precision were 75% and 70%, respectively. The algorithm was unable to disambiguate 25% of the queries due to insufficient statistical co-occurrence data obtained from the monolingual corpus. However, dealing with the sparseness data issue in the revised algorithm, this error rate was reduced by 5%. The finally achieved error rate of 20% was due to the lack of some statistical co-occurrences even after the elimination of the noise term. In addition to this, the ranked translation in the dictionary was not correct for all cases. For example, consider the Arabic query " عجز سداد الدين $\check{g}z$ $sd\bar{a}d$ $\bar{a}ldyn$ " ("The deficient debt payment"). Based on the cohesion score calculated for all possible combinations of the query terms, the term " عجز $\check{g}z$ " has the lowest cohesion score compared to the rest of the terms and thus it is considered to be a noise term and will be eliminated. The rest of the terms " سداد الدين $sd\bar{a}d$ $\bar{a}ldyn$ " have a high enough cohesion score and thus the tool is able to translate them. As discussed in Section 3.1.2, the translation of the eliminated term " عجز $\check{g}z$ " ("deficient") will be selected based on the first ranked translation in the dictionary. The dictionary provided the following translations for the eliminated term: ("rear", "part", "deficit", "insol-

---

[4] www.ldc.upenn.edu/

vency", "incapable", "impotent", "incapacitate", "immobilize", "grow", "old", "weakness" and "inability"). The correct translation of the term "عجز *ǧz* " would be ("deficit"), which is ranked at position number three. The applicability and precision of the proposed algorithm, using the web, averaged over the 20 test queries, were 90% and 85%, respectively. Due to very generic sense, the algorithm was unable to disambiguate 10% of the test queries. For example, consider the Arabic query " رسم جمركي علي اللّوحات الفنية *rasem ǧmrky ⁴lī āllwḥāt ālfnyh* " ("Customs tax of Paintings"). The Arabic word " رسم *rasem* " has the following translations in English, ("drawing", "sketch", "illustration", "fee", "tax", "trace", "indicate", "appoint" and "prescribe"). What made this query very difficult to disambiguate is that the word " رسم *rasem* " can be found frequently in the context of ("Customs") or in the context of ("Paintings"), which both exist in the query. Although the corpus used by the algorithm is rich and covers a broad range of different topics with a significant number of co-occurrence data, this corpus failed to provide co-occurrence data for 25% of the test queries (this error rate is also due to very generic sense cases). In contrast, the algorithm using the co-occurrence data, obtained from the web, could disambiguate 18 queries and failed only to provide co-occurrence data for two queries. This is clearly due to the fact that the web provides significant co-occurrence data compared to other resources.

4.2 Pilot User Study

The goal of the user study was to observe current practice on how real CLIR tasks are accomplished by use of the proposed tool and to imagine a CLIR system that would fully support cross lingual information retrieval tasks. In the performed user study, five users were involved. The types of users are students and researchers. Three of the users were male and two were female. Age ranged from 22 to 32. The differences found between users are more likely to account for the provision of different options to meet more diverse needs. Although it

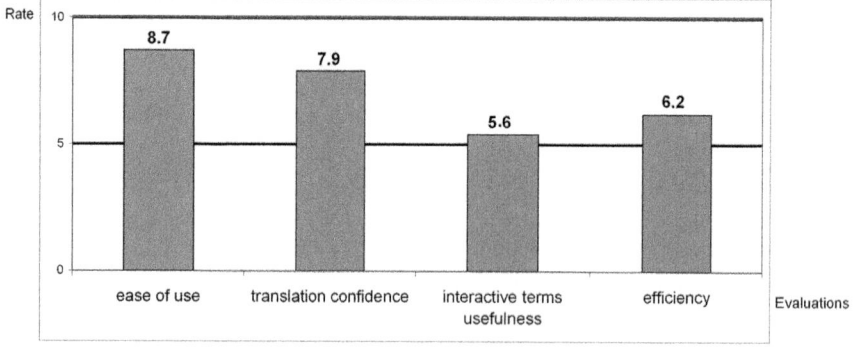

Fig. 3. User questionnaire results for the proposed system

could be argued that the number of users may not be adequate, the strength of this study lies in the fusion of different interests and point of view of the test users, whereby even a single user could counts in building a broad picture of using the proposed tool. Furthermore, according to the research done by Nielsen [14], this small number of test persons is appropriate to find at least 85% of all usability issues. Most of the remaining 15% usability problems we identified by conducting a second user study with a second group of 5 users taking into account the tackling of the previously identified issues. We have chosen this evaluation layout to identify 98% of the possible usability issues in order to ensure that the tool targets the user task as good as possible [13]. At the end of the study and the completion of the tasks, the performance of the system was evaluated by each user, in the form of a questionnaire. This questionnaire included the following question topics to test: ease of use, translation confidence, interactive terms usefulness and efficiency. The users had to give a rating of 0 - 10 (0 - low, 10 - high) for each question. Figure 3 shows the different evaluations that we performed to test the tool:

easy to use addressed the ease of use and was rated with a 8.7. Users found that most system functionalities are very easy to understand and interact with.

Translation confidence addressed how useful and accurate was the contextual information that describes the translation in the source language. The majority of the users found the contextual information which is displayed along with the translation very helpful in giving them a confidence in the translation and rated it with a 7.9. For the improvement request the user suggested decreasing the size of the contextual information (currently, the tool displays 5 documents (sentences) as contextual information). However, decreasing the contextual information size will lead to insufficiency in the interactive terms that can be used to improve the translation. In order to tackle this issue in the future work, and compensated for this insufficiency, we suggest, that the tool can provide a list of interactive terms, regardless of the contextual information.

Interactive terms usefulness addressed the usefulness of the interactive terms in the contextual information that can be used to improve the translation. In many cases the suggested interactive terms by the tool were helpful as the users mentioned. However, in some cases the users needed more terms than the ones suggested by the tool. These terms are color-coded blue and are found in the contextual information, which is displayed along with each proposed translation. Although these terms are found in the context of the user query, these terms in many cases they do not lead to an improvement in the translation therefore the user rating were low 5.6. This issue is interpretable, because currently, we display the contextual information by selecting the most relevant documents to the translation. These documents might have terms which have a very low co-occurrence score with the query in the corpus as a whole. We plan, in future work, to use the contextual information only for translation confidence and provide the user with a list of suggested terms. Currently, the interactive terms are obtained from the contextual information, which are a few documents in size. We plan to use the corpus, as a whole, to obtain these suggested terms. Only terms that have a significant co-occurrence score, with the query terms, will be suggested.

Efficiency addressed how much time a user needed to interact with the tool in order to improve the translation was rated with a 6.2. The required time between submitting the query and receiving the ranked translations along with their contextual information is between 2-5 seconds. Main part of the delay on performing the task is related to the use of araMorph package, that we use to analyze and translate the Arabic query terms. We plan in future work to obtain a full dictionary which we can use to speed up the process of finding possible translations for each query term. This will most likely lead to improved efficiency and accuracy scores.

5   Conclusions

We proposed a context-based CLIR tool, to support the user, in having a certain degree of confidence in the translation. It provides the user with interactive contextual information in order to involve the user in the translation process. Experiments dealing with the accuracy of the tool proved that the tool has a high degree of translation accuracy. Two accuracy experiments were conducted: one using corpus and the other using the Web as a source of the statistical co-occurrences data. For the corpus experiment, the results showed that the performance varied according to the query topics. The translation algorithm was better in the case of topic-specific senses (corpus was built from newswire text mostly concerned with politics and economics) and worse in the case of generic senses. In contrast, using the co-occurrence data obtained from the web improved the accuracy of the translation algorithm. In order to take the user's point of view into account the tool has been tested in form of a user study. The results of the user study were very encouraging, in that the tool could give the users confidence in the translation. Furthermore, the possibility to interactively select term/terms from the contextual information, in order to improve the translation was praised.

References

1. Abdelali, A., Cowie, J.R., Farwell, D., Ogden, W.C.: Uclir: a multilingual information retrieval tool. Inteligencia Artificial, Revista Iberoamericana de Inteligencia Artificial 8(22), 103–110 (2003)
2. Abusalah, M., Tait, J., Oakes, M.: Literature review of cross language information retrieval. World Academy of Science, Engineering and Technology 4, 175–177 (2005)
3. Ahmed, F., Nürnberger, A.: Arabic/english word translations disambiguation using parallel corpus and matching scheme. In: Proceedings of the 12th European Machine Translation Conference (EAMT 2008), pp. 6–11 (2008)
4. Ahmed, F., Nürnberger, A.: Evaluation of n-gram conflation approaches for arabic text retrieval. Journal of the American Society for Information Science and Technology (JASIST) 60(7), 1448–1465 (2009)

5. Aljlayl, M., Frieder, O.: Effective arabic-english cross-language information retrieval via machine readable dictionaries and machine translation. In: Proceedings of the 10th CIKIM Conference (CIKM), pp. 295–302. ACM Press, New York (2001)
6. Ballesteros, L., Croft, B.: Dictionary methods for cross-lingual information retrieval. In: Thoma, H., Wagner, R.R. (eds.) DEXA 1996. LNCS, vol. 1134, pp. 791–801. Springer, Heidelberg (1996)
7. Braschler, M., Peters, C., Schäuble, P.: Cross-language information retrieval (clir) track overview. In: Proceedings of the Eighth Text Retrieval Conference (TREC-8), pp. 25–33 (2000)
8. Clough, P., Stevenson, M.: Cross-language information retrieval using eurowordnet and word sense disambiguation. In: McDonald, S., Tait, J.I. (eds.) ECIR 2004. LNCS, vol. 2997, pp. 327–337. Springer, Heidelberg (2004)
9. Dagan, I., Itai, A.: Word sense disambiguation using a second language monolingual corpus. Computational Linguistics 20(4), 563–596 (1994)
10. Fernandez-Amoros, D., Gil, R.H., Somolinos, J.A.C., Somolinos, C.C.: Automatic word sense disambiguation using cooccurrence and hierarchical information. In: Hopfe, C.J., Rezgui, Y., Métais, E., Preece, A., Li, H. (eds.) NLDB 2010. LNCS, vol. 6177, pp. 60–67. Springer, Heidelberg (2010)
11. Levow, G.-A., Oard, D.W., Resnik, P.: Dictionary-based techniques for cross-language information retrieval. Inf. Process. Manage. 41(3), 523–547 (2005), doi:10.1016/j.ipm.2004.06.012
12. Luca, E.W.D., Hauke, S., Nürnberger, A., Schlechtweg, S.: MultiLexExplorer - combining multilingual web search with multilingual lexical resources. In: Combined Works. on Language-enhanced Educat. Techn. and Devel. and Eval. of Robust Spoken Dialogue Sys., pp. 71–21 (2006)
13. Nielsen, J.: Usability Engineering. Morgan Kaufmann Publishers, San Francisco (1994)
14. Nielsen, J., Landauer, T.K.: A mathematical model of the finding of usability problems. In: CHI 1993: Proceedings of the INTERACT 1993 and CHI 1993 Conference on Human Factors in Computing Systems, pp. 206–213 (1993)
15. Oard, D.W., He, D., Wang, J.: User-assisted query translation for interactive cross-language information retrieval. Information Processing and Management: an International Journal 44(1), 181–211 (2008)
16. Ogden, W.C., Davis, M.W.: Improving cross-language text retrieval with human interactions. In: Proceedings of the 33rd Hawaii International Conference on System Sciences, Washington, DC, USA, p. 3044 (June 2000)
17. Petrelli, D., Beaulieu, M., Sanderson, M., Demetriou, G., Herring, P., Hansen, P.: Observing users, designing clarity: A case study on the user-centered design of a cross-language information retrieval system. Journal of the American Society for Information Science and Technology (JASIST) 55(10), 923–934 (2004)
18. Salton, G.: Experiments in multi-lingual information retrieval. Information Processing Letters 2(1), 6–11 (1973)
19. Shinnou, H., Sasaki, M.: Unsupervised learning of word sense disambiguation rules by estimating an optimum iteration number in the em algorithm. In: Proceedings of the Seventh Conference on Natural Language Learning at HLT-NAACL 2003, pp. 41–48 (2003)
20. Vossen, P. (ed.): EuroWordNet: a multilingual database with lexical semantic networks. Kluwer Academic Publishers, Norwell (1998)

Combining Interaction and Content for Feedback-Based Ranking

Emanuele Di Buccio[1], Massimo Melucci[1], and Dawei Song[2]

[1] University of Padua, Italy
{dibuccio,melo}@dei.unipd.it
[2] The Robert Gordon University, UK
d.song@rgu.ac.uk

Abstract. The paper is concerned with the design and the evaluation of the combination of user interaction and informative content features for implicit and pseudo feedback-based document re-ranking. The features are observed during the visit of the top-ranked documents returned in response to a query. Experiments on a TREC Web test collection have been carried out and the experimental results are illustrated. We report that the effectiveness of the combination of user interaction for implicit feedback depends on whether document re-ranking is on a single-user or a user-group basis. Moreover, the adoption of document re-ranking on a user-group basis can improve pseudo-relevance feedback by providing more effective document for expanding queries.

1 Introduction

Query-based search is the most widespread way to access information. Nevertheless, the intrinsic ambiguity of natural language, query brevity and the personal quality of information needs make query-based search insufficient to deal with every need. It should not come as a surprise that some research works investigate the use and the combination of other approaches to expanding queries by using relevance feedback.

The combination and the exploitation of post-search interaction features ("features" from now on) observed during the interaction between the user and an IR system is one of the most investigated approaches to expanding queries by using relevance feedback. Features are an inexpensive evidence and do not require user effort. They can be gathered by monitoring user behavior when interacting with the documents and may surrogate explicit user's relevance judgments. Hence, the relevance prediction power of features is the main issue.

The gap between what users perceive as relevant to the achievement of an information goal and what IR systems predict to be relevant suggests to exploit personalization at user-system interaction time [1]. However, post-search interaction features are often unavailable, insufficient or unnecessary, thus a broader definition can be useful or necessary. To this end, we consider the evidence gathered from groups of users with similar tasks or requests.

A. Hanbury, A. Rauber, and A.P. de Vries (Eds.): IRFC 2011, LNCS 6653, pp. 46–61, 2011.

This paper is concerned with the combination of interaction and content-based feedback for document re-ranking. We exploit the features of the first documents visited by the user because the users steadily spend less time for searching, examine only the top-ranked results and therefore expect the relevant information in the top-ranked documents [2,3]. We propose a method for extracting behavioral patterns from the features and for representing both retrieved documents and user behavior. Users and groups are sources to distill features. User features are gathered during post-search navigation activity (e.g., when interacting with the results or the landing documents). Group features are distilled from the behavior of the group (e.g., the average dwell time spent on a page). The level of detail present in a set of features is named *granularity*. The specific representation adopted is based on the geometric framework originally proposed in [4]. The basic rationale is to model the behavior of the user when interacting with the first visited documents as a vector subspace. The behavioral patterns extracted from the features observed during the visit are basis vectors and the subspace is that spanned by a subset of these patterns. The subspace is adopted as a new dimension of the information need representation. Each document is represented as a vector of features. Then, documents are matched against the user behavior – the distance between the vector and the subspace is adopted to measure the degree to which a document satisfies the user behavior dimension. Documents are re-ranked according to this distance. In the paper behavior-based representations are adopted to re-rank documents uniquely using the user behavior dimension or to support query modification by extracting terms from the top documents re-ranked by user behavior.

In the paper the P label denotes features distilled from the individual user behavior, namely at personal granularity, while G will denote features at group granularity. In the latter case the value of a feature was obtained as the average computed over all the users that search for that topic other than the user under consideration. Since the adopted framework requires both a representation for the information need (the user behavior dimension) and one for the documents, and two are the possible feature granularities, that leads to four possible combinations X/Y, where X denotes the granularity of the user model and Y that for document representation — X or Y is either P or G.

2   Research Questions

This paper is mainly experimental and aims at addressing the following research questions:

1. When personal data is not available or insufficient, groups of users searching the same topic or performing the same task can be considered as another possible source for features. *What is the effect of the group data on document re-ranking when modeling user behavior and representing documents instead of personal data?* (Section 4.1).
2. Recent research activity has been investigating implicit indicators of relevance. The number of relevant documents affects relevance feedback. Similarly, it affects implicit feedback. *What is the effect of the number of relevant*

documents among those used for user behavior dimension modeling on the effectiveness of document re-ranking? (Section 4.2).

3. If the user behavior-based re-ranking were able to increase the number of good documents for feedback in the top-ranked, query expansion would benefit from re-ranking. Thus, the question is: *What is the effect of top-ranked document re-ordering by user behavior on query expansion?* (Section 4.3).

4. When considering the top-ranked document re-ranked by user behavior dimension as a source for query expansion, a further question is if user behavior-based query expansion is less sensitive to the number of relevant documents among the top-ranked than Pseudo Relevance Feedback (PRF). Thus, the research question is: *What is the effect of the number of relevant documents among the top-ranked on user behavior-based query expansion? Is it less sensitive than PRF?* (Section 4.4)

3   Evaluation

3.1   User Study and Test Collection

The test collection adopted to address the research questions has been created through a user study. Fifteen volunteers have been recruited, particularly three undergraduate students, and twelve among PhD students or postdoctoral researchers. A set of topics was assigned to each user. The users have been asked to examine the top ten retrieved results in response to assigned topics and to assess their relevance with a four-graded scale. Explicit judgments have been gathered through a web application. Moreover, the application monitored the behavior of the user during the assessment, specifically collecting interaction features. The user study has resulted in an experimental dataset which contains content-based document features, interaction features, and explicit judgments of different users on the same document-topic pairs, for a set of topics. The remainder of this section provides a detailed description of the test collection and the experimental tool adopted in the user study, and the collected features.

Test collection for the user study. In the user study we have adopted the Ad-hoc TREC 2001 Web Track Test Collection. The corpus in this test collection is the WT10g, which is constituted by 1,692,096 documents (2.7 GB compressed, 11 GB uncompressed). The test collection includes fifty Ad-hoc topics together with the corresponding relevance judgments[1].

We are interested in comparing the behavior of diverse users when assessing the same topic. However, fifty topics are too many to be judged for each user. Hence, we have considered only a subset of the Ad-hoc topics. When preparing the dataset for the user study, a document has been considered relevant if it is assessed relevant by the TREC assessors. The number of the top-ten retrieved documents that are assessed relevant has been considered as an indicator of *topic difficulty*. The topics without relevant documents, namely 534, 542, 513, 516, and

---

[1] TREC 2001 Web Track data at http://trec.nist.gov/data/t10.web.html

Table 1. TREC 2001 Ad-Hoc web track topics divided according to the number of relevant documents in the top 10 retrieved

Difficulty	Number of relevant documents	Topics
High	1-2	506-517-518-543-546
Medium	3-5	501-502-504-536-550
Low	6-10	509-510-511-544-549

Table 2. Topic sets, each one constituted by three topics for each set in Table 1

	Difficulty		
Topic Set	High (1-2)	Medium (3-5)	Low (6-10)
A	506-517-518	501-502-504	509-510-511
B	517-518-543	502-504-536	510-511-544
C	518-543-546	504-536-550	511-544-549

531, have been removed. The remaining topics have been divided by difficulty in three sets according to the number of relevant documents. Five topics have been randomly selected from each difficulty level, thus obtaining the fifteen topics reported in Table 1. Three distinct sets of nine queries have been built, each set being composed of three topics for each set of Table 1, thus achieving the Latin squares reported in Table 2. We have decided to distribute the topics so that at least one topic from each set would be assessed by all the users and the average topic difficulty was uniform per user. One of the three sets of topics, namely A, B, or C, was assigned to each user.

The use of a test collection is crucial since it allows us to simulate a realistic scenario in which the system is unaware of the real user's information need and it has to exploit the description provided by the topic. The document corpus has been indexed by the Indri Search Engine[2]. English stop-words have been removed and the Porter stemmer has been adopted. The documents of the WT10g have been ranked by Indri (default parameters were used) and the top 10 documents have been considered for each topic.

Experimental tool. We have developed a web application to collect the information about the user interaction behavior. The first web page presented to the user provided the list of topic identifiers. Once the user selected an assigned topic, a new web page divided in three frames was presented. The upper right frame reports the topic descriptions, i.e. title, narrative, and description; the left frame reports the title of the top ten retrieved documents ranked by Indri. A user reads the document in the bottom right frame; the user could access the documents in any order. A drop down menu allows the user to select the relevance degree

---

[2] http://www.lemurproject.org/indri/

Table 3. Features adopted to model the *user behavior* dimension and to represent documents

Feature	Description
Features observed from document/browser window	
query terms	number of topic terms displayed in the title of the corresponding result
ddepth	depth of the browser window when examining the document
dwidth	width of the browser window when examining the document
doc-length	length of the document (number of terms)
Features observed from the user behavior	
display-time	time the user spent on the page in his first visit
scroll-down	number of actions to scroll down the document performed both by page-down and mouse scroll
scroll-up	number of actions to scroll up the document performed both by page-up and mouse scroll
sdepth	maximum depth of the page achieved by scrolling down, starting from the ddepth value

of the document. We have adopted the four graded relevance scale – (0) non relevant, (1) marginally, (2) fairly, (3) highly relevant – proposed in [5].

Features. Each action concerning the selection of the topic, the selection of a result and the relevance assessments have been centrally stored; both information about the type of action and the timestamp have been collected. Other features have been stored locally in the browser cookies. The features gathered from the user study are reported in Table 3. They can be divided in two groups: the features concerning the results or the displayed document, specifically the way in which they have been presented, and the features concerning user behavior. Document length was considered together with the display-time because a large display-time on a short document can have a different meaning than a display-time on a long document. The dimensions of the browser window have been considered together with the scrolling actions because different styles of scrolling interactions observed for diverse users can be also due to the different size of the browser window when visiting the same document with regard to the same query.

At the end of the evaluation session, the file with the cookies stored by the browser where the interaction data have been stored, was returned by each participant. Two users did not assess all the documents in the result list for some topics. For this reason, only the user behavior of thirteen among the fifteen users have been considered in this work, for a total of 79 (user,topic) pairs and 790 entries where each entry refers to the visit of a specific user to a particular document with regard to a topic.

3.2   Re-ranking Methods

User behavior-based re-ranking. A user visited n documents among the ten displayed in the result page returned in response to a query, the latter list (provided by Indri) being the *baseline*. Thus, for each query q and for each user u who searched using that query, namely for each pair (q,u), the following steps have been performed:

1. *Selection of the combination of the source for features.* Either P/P, P/G, G/P or G/G has been selected.
2. *Collection of the features from the first $n_B = 3$ visited documents.* The collected features are prepared in a $n_B \times k$ matrix where k is the number of features collected from the n_B visited documents. The reason for adopting the top visited documents and not the top ranked documents is to simulate a scenario where the first data obtained from interaction are adopted for feedback; the first visited results observed during the user study differ from the top ranked results. When considering the G source the value of a feature is obtained as the average computed over all the users that search for that topic other than the user under consideration.
3. *Modeling user behavior dimension by using patterns.* We applied Principal Component Analysis (PCA) [6] on the $n_B \times k$ matrix. The result of PCA is an orthonormal basis such that the basis vectors are 1:1 correspondence to the patterns. Thus, a pattern **p** is an eigenvector associated with non-null eigenvalue.
4. *Representation of the top-ranked documents.* Each document is represented as a vector **y** of the features reported in Table 3. Feature values are distilled from the source Y of the combination X/Y selected at step 1.
5. *Re-ranking of the top-ranked documents.* Each feature of **y** is used to retrieve the documents associated to the feature and ranked on the top of the initial list by the system. For all patterns, $|\mathbf{y}'\mathbf{p}|^2$ is computed using a document-at-a-time-like algorithm. The best performing pattern is manually selected.
6. *Effectiveness measurement.* The NDCG@n (for different n's) is computed for the new result list obtained after document re-ranking. DCG is computed according to the alternative formulation reported in [7], namely

$$DCG(i) = \sum_i (2^{r(i)} - 1)/\log(i + 1),$$

where $r(i)$ is the relevance of the document at position i. The normalization factor is the DCG of the perfect ranking. The gains adopted to compute NDCG are those provided by the user u when assessing the query q.

User behavior-based re-ranking to support query expansion. Besides the impact on document re-ranking, the effectiveness of user behavior to support query expansion is investigated. It is supposed that a first stage prediction has been performed based on Indri. For each query the following steps are performed:

1. Consider the top $n_B = 3$ documents retrieved by the baseline.
2. Perform step 1–5 described in the previous section by the G/G combination for user behavior dimension-based re-ranking using the $n_B = 3$ documents considered in the previous step. Consider the top $n_F = 5$ documents re-ranked by user behavior dimension.
3. Re-ranking of the top $m = 50$ documents returned by the baseline by using the PRF algorithm of Indri, adaptation of relevance models proposed in [8], on the $n_F = 5$ considered documents with $k = 10$ expansion terms.
4. Computation of the NDCG@n (for different n's) for the new result list obtained from the feedback at step 3.

This user-behavior based query expansion (IRF) is compared with the PRF algorithm of Indri on the top $n_F = 5$ documents retrieved by the baseline.

At step 2, the combination of sources of features adopted is G/G, that is the tests are performed in a non personalized scenario. Interaction features of all the users who searched using the query are adopted for dimension and document modeling for user behavior-based re-ranking. The dimension is automatically obtained using the first eigenvector among those extracted.

At step 3, when considering IRF the strategy is not actually PRF since we are using the top re-ranked by user behavior dimension.

Differently from the previous re-ranking method (user dimension-based re-ranking without query expansion), the gains adopted in Step 4 for the computation of the NDCG@n are those provided by the TREC assessors, that is those in the qrels of the TREC 2001 Web Track Test Collection. The basic idea is that the TREC assessor is considered as a new user, not among those in the group, who will be supported using group evidence. In other words this experiment aims at investigating if the pattern extracted by PCA could be useful for a non-personalized re-ranking.

4   Experimental Results

4.1   Question 1: Effect of Group Data on Document Re-ranking

The first research question concerns with the impact of the selection of the source combination on document re-ranking (i.e., P/P vs. G/- and -/G). In particular, we are interested in understanding if using group data both for modeling the user behavior and for representing documents negatively affects document re-ranking in comparison with exploiting the data distilled from the individual. Some preliminary results for this research question only for $n_B = 3$ have been reported in [9]; $n_B < 3$ has not been considered because the number of patterns for the diverse combinations was usually one and this pattern was not effective. Here we extend the analysis with varying n_B.

Table 4 reports the mean and the median NDCG@10 over all the (topic,user) pairs for all the combinations when $3 \leq n_B \leq 10$ and the best performing pattern is considered for each pair. Looking at the mean and the median NDCG@10, the results show that P/P and G/G benefit from additional evidence, yet NDCG@10

Table 4. Comparison among median NDCG@10 of the diverse source combinations when varying the number of documents n_B used to obtain the user behavior dimension. Values marked by asterisks are those for which the difference with the P/P case was significant (one asterisk denotes $p < 0.05$, two asterisks $p < 0.01$) according to the one-sided Wilcoxon signed ranked test based on the alternative hypothesis that P/P combination performed better than the group-based combinations.

	Personal	Group			Increment (%)		
n_B	P/P	P/G	G/P	G/G	$\Delta_{P/G}$	$\Delta_{G/P}$	$\Delta_{G/G}$
3	0.817	0.832	0.799	0.825	1.748	-2.238	0.922
4	0.839	0.825	0.805*	0.827	-1.615	-4.056	-1.324
5	0.833	0.835	0.826	0.835	0.288	-0.817	0.288
6	0.839	0.843	0.833	0.839	0.524	-0.656	0.045
7	0.847	0.840	0.831	0.848	-0.798	-1.829	0.137
8	0.841	0.832	0.839**	0.835	-1.164	-0.333	-0.701
9	0.847	0.835	0.839**	0.838	-1.351	-0.985	-1.109
10	0.853	0.835	0.839**	0.832	-2.049	-1.686	-2.506

increased monotonically with the number of documents used as evidence for none of the combinations. The results obtained from the different combinations are comparable. The only significant difference is observed for the G/P case, whose poor performance could be due both to the fact the a non-personalized dimension is adopted and the comparison is performed between a dimension and a document representation obtained from diverse sources for interaction features. Because of its lack of effectiveness, in the remainder of this work this combination will be no longer considered.

Moreover, results in Table 4 shows that, even if the re-ranking effectiveness is comparable, the adoption of group data can negatively affect re-ranking. Indeed, even if the G/G combination seems to be promising for $n_B = 3$, when considering the results per topic and per user they show that also in this case the adoption of group data can affect effectiveness of re-ranking. This is the case, for instance, of the topics 536, 543 and 550 where all the three combinations involving group data performed worse than the P/P case.

The above remarks concerned with the comparison among the diverse combinations thus investigating the effect of using group data instead of personal data for personalized user behavior-based re-ranking. But no comparison was performed with the baseline B. Table 5 reports the median NDCG@10 for the baseline and the diverse combinations for different values of n_B. None of the combinations significantly outperformed the baseline ranking. Also when comparing results per topic and per user none of the combinations outperformed the baseline for all the topics or all the users.

The relatively small number of experimental records is definitely a limitation since small numbers make the detection of significant differences harder than the detection based on large datasets. But, the small size of the dataset allows us to note that the two sources of features (i.e. P and G) adopted seem to provide

Table 5. Comparison among median NDCG@10 of the baseline and the diverse source combinations when varying the number of documents n_B used to obtain the user behavior dimension

	Baseline	Source combinations			Increment (%)		
n_B	B	P/P	P/G	G/G	$\Delta_{P/P-B}$	$\Delta_{P/G-B}$	$\Delta_{G/G-B}$
3	0.838	0.817	0.832	0.825	-2.462	-0.757	-1.563
4	0.838	0.839	0.825	0.827	0.053	-1.563	-1.272
5	0.838	0.833	0.835	0.835	-0.604	-0.317	-0.317
6	0.838	0.839	0.843	0.839	0.053	0.577	0.098
7	0.838	0.847	0.840	0.848	1.048	0.242	1.187
8	0.838	0.841	0.832	0.835	0.387	-0.781	-0.317
9	0.838	0.847	0.835	0.838	1.048	-0.317	-0.072
10	0.838	0.853	0.835	0.832	1.769	-0.317	-0.781

diverse contributions. For instance, for topics 518, 536, and 546 only one of the two combinations performs better than the baseline. This suggests to investigate combinations of the diverse feature granularities.

4.2   Question 2: Effect of the Number of Relevant Documents on Document Re-ranking

The representation of the user behavior exploits the data gathered from the first visited documents by the users, extracts possible patterns (i.e. eigenvectors of the correlation matrix) from those data and uses the most effective pattern for re-ranking. If the visited documents are relevant, it is necessary to investigate whether the improvement in terms of effectiveness can mainly be due to the ability of the user to select relevant documents. To this end, we investigated the relationship between the number of relevant documents among the top $n_B = 3$ visited and NDCG@10 across the diverse combinations.

In Figure 1 the results are depicted. The NDCG@10's measured for the baseline (`Indri`) when considering all the users and all the topics is plotted against the number of relevant documents in the top three visited — the regression lines are reported for providing an idea of the trend. The least steep lines refer to P/P, P/G and G/G. For the diverse combinations the dependence with the number of relevant documents among the top three visited is still linear, but the slope decreases and the intercept increases.

The mean and the median NDCG@10 was higher than the baseline when only one relevant document was present among those used for feedback, but this increment decreased when increasing the number of relevant documents; the same results have been observed for $n_B \in \{4, 5\}$. The main limitation is the robustness of the adopted approach. Indeed, when observing the variance of NDCG@10 values, in the event of one relevant document the variance is smaller than those obtained for the baseline; differently, when the number of relevant

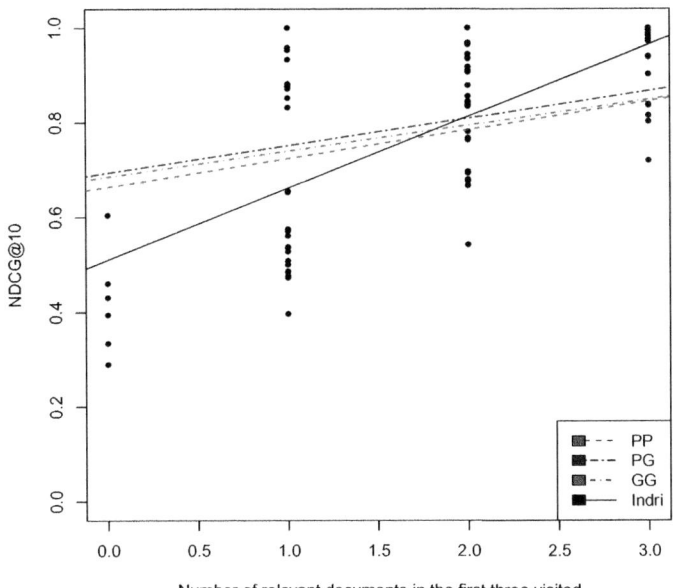

Fig. 1. Comparison among the regression line of the baseline(`Indri`) and those of the diverse combinations (`X/Y`'s)

documents increases, the baseline has smaller variance, thus suggesting that even if the user behavior based re-ranking can provide some improvement, the latter is less robust than the baseline.

4.3 Question 3: Effect of User Behavior-Based Document Re-ranking on Query Expansion

The results reported above showed that the improvement in terms of retrieval effectiveness is not consistent throughout all the topics or all the users, but the effectiveness of the top ten document re-ranking seems to be not strictly dependent from the relevance of the document used to model the user behavior. For this reason we investigated if the top n_F documents re-ranked by the user behavior are a more effective evidence for query expansion than the top n_F retrieved by the baseline. The basic idea is to investigate if user behavior-based re-ranking is able to bring at high rank positions good sources for query expansion, thus improving the effectiveness respect to PRF where the top ranked documents retrieved by the baseline are supposed to be good sources for feedback.

Table 6 reports the mean and the median NDCG@n's computed over the different values of the parameters (k, n_F) adopted, that is for diverse number of expansion terms and feedback documents; the number of documents used to model the user behavior dimension was fixed to $n_B = 3$, since it provided good results also for group-based source combinations. The results show that query

Table 6. Median and Mean NDCG@n, with $n \in \{10, 20, 30, 50\}$, computed over all the values of the parameters k and n_F, where k denotes the number of expansion terms and varies in $\{5, 10, 15, 20, 25, 30\}$, and n_F denotes the number of feedback documents and varies in $\{1, 2, 3, 4, 5\}$. Results refer to the case where $n_B = 3$.

	NDCG@10			Δ (%)		NDCG@20			Δ (%)	
	B	PRF	IRF	$\Delta_{\text{PRF-B}}$	$\Delta_{\text{IRF-B}}$	B	PRF	IRF	$\Delta_{\text{PRF-B}}$	$\Delta_{\text{IRF-B}}$
median	0.324	0.319	0.341	-1.31	5.44	0.295	0.286	0.311	-3.15	5.23
mean	0.329	0.318	0.350	-3.45	6.38	0.295	0.290	0.312	-1.69	5.60

	NDCG@30			Δ (%)		NDCG@50			Δ (%)	
	B	PRF	IRF	$\Delta_{\text{PRF-B}}$	$\Delta_{\text{IRF-B}}$	B	PRF	IRF	$\Delta_{\text{PRF-B}}$	$\Delta_{\text{IRF-B}}$
median	0.249	0.281	0.293	12.86	17.78	0.208	0.207	0.220	-0.48	5.80
mean	0.288	0.285	0.303	-0.95	5.05	0.228	0.217	0.225	-5.10	-1.35

expansion can benefit from user behavior based re-ranking, even if the improvements are modest: except for NDCG@30, the improvement is approximately 5%. Table 7 reports the NDCG@n's for different values of k, namely the number of expansion terms, and different n_F's, namely the number of feedback documents. The results show that the adopted approach, Implicit Relevance Feedback (IRF), benefits from a small number of feedback documents, $n_F = \{1, 2\}$, and an increment of the number of terms used for query expansion, i.e. $k = \{10, 20\}$. For most of the parameters pairs (k, n_F) IRF can improve PRF. But this specific approach should be improved since it is not robust. Let us consider, for instance, the case for $k = 10$ and $n_F = 2$, where PRF did not improve the baseline ($\Delta_{\text{PRF-B}} = 0.24\%$) differently from IRF ($\Delta_{\text{IRF-B}} = 8.67\%$), and the difference in terms of NDCG@10 was greater than 5%. Table 8 reports the results for each topic and shows that also in this case, IRF is not able to outperform PRF for all the topics.

The user behavior-based re-ranking is able to provide an improvement respect to PRF, increasing the number of good sources for feedback at high rank positions and supporting feedback when small evidence is adopted as input, e.g. one or two documents. As shown in Table 7, for 39/45 cases IRF performed as good as or better than PRF, and for 26/45 the increment in terms of NDCG was higher than 5%. But an analysis of the effectiveness per topic shows that a more robust approach is required since for some topics PRF performed better than IRF.

4.4 Question 4: Effect of the Number of Relevant Documents Used for Dimension Modeling on Query Expansion

The objective of the research question discussed in Section 4.3 was to investigate the capability of the user behavior dimension to increase the number of good sources, namely documents, for query expansion at high rank position, thus increasing the effectiveness of Pseudo Relevance Feedback. The results showed that PRF can benefit from a preliminary user behavior based re-ranking. In order to

Table 7. Comparison among the NDCG@n's of the baseline (B), PRF and IRF for different values of n, n_F and k. The results in bold type are those for which the increment respect to the baseline B is higher than 5%. The results marked by an asterisk are those for which the increment of IRF respect to PRF in terms of NDCG@n, $\Delta_{\text{IRF}-\text{PRF}}$, is greater than 5%; those marked by two asterisks are those for which $\Delta_{\text{IRF}-\text{PRF}} > 10\%$.

k	n_F	NDCG@10			NDCG@20			NDCG@30		
		B	PRF	IRF	B	PRF	IRF	B	PRF	IRF
5	1	0.329	0.341	0.343	0.295	0.301	0.308	0.288	0.290	0.296
	2	0.329	**0.348**	0.313	0.295	0.309	0.287	0.288	0.296	0.282
	3	0.329	0.302	0.320*	0.295	0.287	0.293	0.288	0.283	0.289
	4	0.329	0.340	0.331	0.295	0.305	0.297	0.288	0.295	0.293
	5	0.329	0.299	0.312	0.295	0.278	0.292*	0.288	0.277	0.284
10	1	0.329	**0.351**	**0.370***	0.295	**0.313**	0.326	0.288	0.298	**0.309**
	2	0.329	0.330	**0.357****	0.295	0.291	**0.317***	0.288	0.285	**0.305***
	3	0.329	0.292	0.342**	0.295	0.283	0.303*	0.288	0.280	0.300*
	4	0.329	0.286	0.338*	0.295	0.269	0.301**	0.288	0.273	0.300*
	5	0.329	0.311	0.324	0.295	0.283	0.293	0.288	0.278	0.295*
20	1	0.329	**0.371**	**0.378**	0.295	**0.313**	**0.331***	0.288	**0.305**	**0.319**
	2	0.329	0.328	**0.376****	0.295	0.292	**0.324****	0.288	0.286	**0.311***
	3	0.329	0.280	0.341**	0.295	0.277	**0.315****	0.288	0.278	0.299*
	4	0.329	0.299	0.334**	0.295	0.274	0.309**	0.288	0.280	0.298*
	5	0.329	0.288	0.343*	0.295	0.273	0.299*	0.288	0.277	0.298*

Table 8. NDCG@$\{10, 20, 30\}$'s per topic for IRF and PRF when $k = 10$ and $n_F = 2$

Topic	NDCG@10			NDCG@20			NDCG@30		
	PRF	IRF	$\Delta(\%)$	PRF	IRF	$\Delta(\%)$	PRF	IRF	$\Delta(\%)$
501	0.386	0.432	12.02	0.400	0.497	24.22	0.408	0.505	23.82
502	0.300	0.454	51.62	0.219	0.318	45.68	0.199	0.276	38.62
504	0.393	0.333	-15.10	0.316	0.289	-8.52	0.372	0.308	-17.11
506	0.125	0.108	-13.84	0.125	0.108	-13.84	0.125	0.108	-13.84
509	0.518	0.518	0.00	0.478	0.478	0.00	0.466	0.466	0.00
510	0.697	0.697	0.00	0.450	0.450	0.00	0.393	0.393	0.00
511	0.310	0.323	4.36	0.337	0.383	13.57	0.370	0.395	6.62
517	0.140	0.155	10.81	0.154	0.121	-21.40	0.157	0.139	-11.43
518	0.000	0.000	-	0.102	0.070	-31.96	0.101	0.100	-1.19
536	0.355	0.426	20.12	0.269	0.323	20.13	0.291	0.345	18.68
543	0.064	0.078	23.27	0.041	0.051	23.11	0.037	0.045	23.29
544	0.673	0.700	4.10	0.670	0.737	9.97	0.633	0.675	6.67
546	0.169	0.240	42.03	0.188	0.266	41.48	0.190	0.242	27.37
550	0.489	0.539	10.25	0.328	0.348	6.04	0.252	0.267	6.03
all	0.330	0.357	8.40	0.291	0.317	8.86	0.285	0.305	6.80

Table 9. Median NDCG@$\{10, 20, 30\}$'s for different numbers of relevant documents among the top three documents of the baseline, when considering $n_F = 3$. In the event of PRF, this number corresponds to the number of relevant documents among those used for feedback. In the event of IRF, this number corresponds to the number of relevant documents among those used for modeling the user behavior dimension.

Relevant in Top 3	NDCG@10			NDCG@20			NDCG@30		
	PRF	IRF	$\Delta(\%)$	PRF	IRF	$\Delta(\%)$	PRF	IRF	$\Delta(\%)$
0	0.074	0.069	-6.09	0.067	0.045	-33.13	0.075	0.076	1.27
1	0.221	0.252	13.95	0.251	0.288	14.84	0.240	0.256	6.72
2	0.342	0.454	32.71	0.367	0.387	5.28	0.341	0.406	19.00
3	0.514	0.514	0.11	0.475	0.475	-0.09	0.472	0.497	5.24

Table 10. Median NDCG@$\{10, 20\}$'s (Table 10a) and NDCG@30's (Table 10b) for different numbers of relevant documents among the top 3 of the baseline. Results are reported for the baseline and the two feedback strategies IRF and PRF.

Relevant in Top 3	NDCG@10					NDCG@20				
	B	PRF	IRF	$\Delta_{\text{PRF-B}}$	$\Delta_{\text{IRF-B}}$	B	PRF	IRF	$\Delta_{\text{PRF-B}}$	$\Delta_{\text{IRF-B}}$
0	0.069	0.074	0.069	6.48	-6.09	0.045	0.067	0.045	49.55	0.00
1	0.166	0.221	0.252	33.04	13.95	0.187	0.251	0.288	33.91	53.78
2	0.437	0.342	0.454	-21.78	32.71	0.382	0.367	0.387	-3.82	1.26
3	0.625	0.514	0.514	-17.88	0.11	0.596	0.475	0.475	-20.28	-20.35

(a)

Relevant in Top 3	NDCG@30				
	B	PRF	IRF	$\Delta_{\text{PRF-B}}$	$\Delta_{\text{IRF-B}}$
0	0.040	0.075	0.076	87.56	89.95
1	0.162	0.240	0.256	48.65	58.65
2	0.381	0.341	0.406	-10.46	6.56
3	0.591	0.472	0.497	-20.12	-15.93

(b)

gain more insights into the user behavior dimension capability to support query expansion, we investigated the effect of the number of relevant documents among the top n_B. The objective is to understand if, also when there is a small number of relevant documents among the top n_B, actually those adopted to model the dimension, user behavior-based re-ranking is able to improve the effectiveness of the system in ranking highly relevant documents at high rank positions.

Table 9 reports the NDCG@n's for different values of n and different numbers of relevant documents among the top three documents of the baseline, when considering $n_F = 3$. In the event of PRF, this number corresponds to the number of relevant documents among those used for feedback. In the event of IRF, this number corresponds to the number of relevant documents among those used for modeling the user behavior dimension. When there are no relevant documents among the top 3 of the baseline the effectiveness of feedback is low and PRF performs better; this results suggests that when no relevant documents are adopted for dimension modeling the effectiveness of the model is negatively affected. Differently when only one or two relevant documents are present in the top 3 used for pseudo-feedback (PRF) or dimension modeling (IRF), IRF outperforms PRF thus suggesting that is able to improve the number of good sources for content-based feedback in the top 3. When the number of relevant documents is three, namely all the feedback documents are relevant, the two approaches perform equally.

Tables 10 reports the median NDCG@n's for the two feedback strategies when compared to the baseline. IRF was able to provide a positive contribution when one or two relevant documents are present in the top 3, but both the feedback techniques hurt the initial ranking when the top three documents are relevant.

5   Related Work

A review on past works investigating implicit indicators and feedback techniques is reported in [10]. In that work individual and group granularities referred to two distinct dimensions for classification: individual's and group granularity levels refer to explicit judgments that the implicit feedback strategy should predict.

Collaborative filtering, for instance, exploits group ratings gathered by similar users to predict the user interests. An application to web search that involves interaction data at group granularity is [11] where the author investigated the predicting effectiveness of click-through data gathered from users in a community, e.g. a interest-specific web portal. In [12] diverse grouping criteria are investigated for tag recommendation in a social network scenario. Users are grouped according to explicit connections with other members or according to the subscription to interest groups. Tag occurrence and co-occurrence information at personal and group levels are adopted to estimate the probability for ranking tags to suggest. Even if the work on collaborative filtering and recommendation is related, this paper concentrates on IRF and PRF as well as on the "tension" between users with the same topic in mind. The work reported in [13] is also related but it is focused on different criteria for group creation and the proposed *groupization* algorithm consists in aggregating personalized scores.

In regard to implicit indicators granularities, in [14] the authors investigated combination of implicit indicators by Bayesian modeling techniques. Two feature granularities have been considered: result-level features which referred to individual pages visited by the user for a specific query, and session-level features whose value, when the features are not session specific, was obtained as the average value of result-level features computed over all the queries in the session. In [15] group granularity interaction feature values are adopted together with derived features to learn user models. The value of a derived feature was obtained subtracting the feature background distribution value from the observed value: the assumption is that the relevance component of a feature can be obtained by considering its deviation from its background distribution. The value for a feature at group granularity was obtained as the average value computed across all the users and search sessions for each query-URL pair.

Also in this paper group granularity features are obtained as the average computed over all the users for a specific topic, but not including the user the IR system aims at supporting. Differently from [15] this paper is focused on the capability of indicators to support personalization and our approach is based on a different hypothesis: relevance information can be extracted from the correlation among the observed indicators. The approach adopted is that proposed in [16] where PCA was used to extract behavioral patterns, then used for document re-ranking; re-ranked documents are then adopted as source for query expansion. Differently from [16] this work investigates the impact of source combinations on the re-ranking effectiveness, and the effectiveness of group behavior document re-ranking to support query expansion in a non-personalized search task.

6   Conclusion

The results reported in this paper show that the contributions of the diverse sources are comparable, thus making personalized IRF feasible despite the data sparsity observed when the interaction features are collected on a per-user basis. Another finding was that the diverse source combinations X/Y's provide complementary contributions with respect to the baseline for some topics and users, thus suggesting to investigate source combinations when both of them are available. Moreover, the effectiveness of query expansion based on the highest ranking results re-ranked by group behavior was investigated in a non-personalized search task. The results also show that the highest ranking results re-ranked by group behavior are comparable with the highest ranking results in the baseline list when used for query expansion, thus suggesting to investigate combinations both of content and user behavior as evidence to support query expansion at a larger scale than the study of this paper to see whether they can effectively complement each other or whether they instead tend to cancel each other out.

Acknowledgements

The work reported in this paper has been supported by the PROMISE network of excellence (contract n. 258191) project and by the QONTEXT project (grant agreement n. 247590), as a part of the 7th Framework Program of the European commission.

References

1. Teevan, J., Dumais, S.T., Horvitz, E.: Potential for personalization. ACM Transactions on Computer-Human Interaction 17, 4:1–4:31 (2010), http://doi.acm.org/10.1145/1721831.1721835, doi:10.1145/1721831.1721835

2. Jansen, B.J., Spink, A.: An analysis of web searching by european alltheweb.com users. Information Processing and Management 41, 361–381 (2005), doi:10.1016/S0306-4573(03)00067-0

3. Jansen, B.J., Spink, A.: How are we searching the world wide web?: a comparison of nine search engine transaction logs. Information Processing and Management 42, 248–263 (2006), http://dx.doi.org/10.1016/j.ipm.2004.10.007, doi:10.1016/j.ipm.2004.10.007

4. Melucci, M.: A basis for information retrieval in context. ACM Transaction on Information Systems 26, 14:1–14:41 (2008), http://doi.acm.org/10.1145/1361684.1361687, doi:10.1145/1361684.1361687

5. Järvelin, K., Kekäläinen, J.: Cumulated gain-based evaluation of IR techniques. ACM Transactions on Information Systems 20, 422–446 (2002), http://doi.acm.org/10.1145/582415.582418, doi:10.1145/582415.582418

6. Pearson, K.: On lines and planes of closest fit to systems of points in space. Philosophical Magazine 2, 559–572 (1901)

7. Croft, B., Metzler, D., Strohman, T.: Search Engines: Information Retrieval in Practice, 1st edn. Addison-Wesley Publishing Company, USA (2009)

8. Lavrenko, V., Croft, W.B.: Relevance based language models. In: Proceedings of SIGIR 2001, New Orleans, Louisiana, United States, pp. 120–127. ACM, New York (2001), http://doi.acm.org/10.1145/383952.383972, doi:10.1145/383952.383972

9. Di Buccio, E., Melucci, M., Song, D.: Exploring Combinations of Sources for Interaction Features for Document Re-ranking. In: Proceedings of HCIR 2010, New Brunswick, NJ, USA, pp. 63–66 (2010)

10. Kelly, D., Teevan, J.: Implicit feedback for inferring user preference: a bibliography. SIGIR Forum 37, 18–28 (2003), http://doi.acm.org/10.1145/959258.959260, doi:10.1145/959258.959260

11. Smyth, B.: A community-based approach to personalizing web search. Computer 40, 42–50 (2007), doi:10.1109/MC.2007.259

12. Rae, A., Sigurbjörnsson, B., van Zwol, R.: Improving tag recommendation using social networks. In: Proceedings of RIAO 2010, Paris, France, pp. 92–99 (2010)

13. Teevan, J., Morris, M.R., Bush, S.: Discovering and using groups to improve personalized search. In: Proceedings of WSDM 2009, Barcelona, Spain, pp. 15–24. ACM, New York (2009), http://doi.acm.org/10.1145/1498759.1498786, doi:10.1145/1498759.1498786

14. Fox, S., Karnawat, K., Mydland, M., Dumais, S., White, T.: Evaluating implicit measures to improve web search. ACM Transactions on Information Systems 23, 147–168 (2005), http://doi.acm.org/10.1145/1059981.1059982, doi:10.1145/1059981.1059982

15. Agichtein, E., Brill, E., Dumais, S., Ragno, R.: Learning user interaction models for predicting web search result preferences. In: Proceedings of SIGIR 2006, Seattle, Washington, USA, pp. 3–10. ACM, New York (2006), http://doi.acm.org/10.1145/1148170.1148175, doi:10.1145/1148170.1148175

16. Melucci, M., White, R.W.: Utilizing a geometry of context for enhanced implicit feedback. In: Proceedings of CIKM 2007, Lisbon, Portugal, pp. 273–282. ACM, New York (2007), http://doi.acm.org/10.1145/1321440.1321480, doi:10.1145/1321440.1321480

Query Expansion for Language Modeling Using Sentence Similarities

Debasis Ganguly, Johannes Leveling, and Gareth J.F. Jones

CNGL, School of Computing, Dublin City University, Ireland
{dganguly,jleveling,gjones}@computing.dcu.ie

Abstract. We propose a novel method of query expansion for Language Modeling (LM) in Information Retrieval (IR) based on the similarity of the query with
sentences in the top ranked documents from an initial retrieval run. In justification
of our approach, we argue that the terms in the expanded query obtained by the
proposed method roughly follow a Dirichlet distribution which, being the conjugate prior of the multinomial distribution used in the LM retrieval model, helps
the feedback step. IR experiments on the TREC ad-hoc retrieval test collections
using the sentence based query expansion (SBQE) show a significant increase in
Mean Average Precision (MAP) compared to baselines obtained using standard
term-based query expansion using LM selection score and the Relevance Model
(RLM). The proposed approach to query expansion for LM increases the likelihood of generation of the pseudo-relevant documents by adding sentences with
maximum term overlap with the query sentences for each top ranked pseudo-
relevant document thus making the query look more like these documents. A per
topic analysis shows that the new method hurts less queries compared to the baseline feedback methods, and improves average precision (AP) over a broad range
of queries ranging from easy to difficult in terms of the initial retrieval AP. We
also show that the new method is able to add a higher number of *good* feedback
terms (the golden standard of *good* terms being the set of terms added by True
Relevance Feedback). Additional experiments on the challenging search topics of
the TREC-2004 Robust track show that the new method is able to improve MAP
by 5.7% without the use of external resources and query hardness prediction typically used for these topics.

1 Introduction

A major problem in information retrieval (IR) is the mismatch between query terms
and terms in relevant documents in the collection. Query expansion (QE) is a popular
technique used to help bridge this vocabulary gap by adding terms to the original query.
The expanded query is presumed to better describe the information need by including
additional or attractive terms likely to occur in relevant documents.

Evidence suggests that in some cases a document as a whole might not be relevant
to the query, but a subtopic of it may be highly relevant [1], summarization improves
accuracy and speed of user relevance judgments [2], and even relevant documents may
act as "poison pills" and harm topic precision after feedback [3].

This problem is compounded in Blind Relevance Feedback (BRF) where all top
ranked documents from an initial retrieval run are assumed to be relevant, but often

A. Hanbury, A. Rauber, and A.P. de Vries (Eds.): IRFC 2011, LNCS 6653, pp. 62–77, 2011.

one or more of them will not be. In this case, terms not related to the topic of the query, but which meet the selection criteria are used for QE (e.g. semantically unrelated, but high frequency terms from long pseudo-relevant documents). Using text passages for feedback (sentences instead of full documents) has the potential to be more successful since long documents can contain a wide range of discourse and using whole documents can result in noisy terms being added to the original query, causing a topic shift.

The multi-topic nature of many long documents means that the relevant portion may be quite small, while generally feedback terms should be extracted from relevant portions only. This observation leads to the idea of using smaller textual units (passages[1]) for QE. This proposal dates back to early experiments on the TIPSTER collections [4,5]. This approach raises questions of how to create the passages, how to select the relevant passages, how passage size influences performance, and how to extract feedback terms from the passages. In the method proposed in this paper, passages correspond to single sentences, which are ranked by their similarity with the original query. Using sentences instead of fixed length (non)overlapping word windows has the implicit advantages that firstly it does not involve choosing the window length parameter, and secondly a sentence represents a more natural semantic unit of text than a passage.

The sentence based QE method for LM proposed in this paper is based on extracting sentences from pseudo-relevant documents, ranking them by similarity with the query sentences, and adding the most similar ones as a whole to expand the original query. This approach to QE, which we call SBQE (Sentence Based Query Expansion), introduces more context to the query than term based expansion. Moreover adding a number of sentences from the document text in its original form has a natural interpretation within the context of the underlying principle of LM in the sense that the modified query more closely resembles the pseudo-relevant documents increasing the likelihood of generating it from these documents.

The remainder of the paper is organized as follows: Section 2 reviews existing and related work on BRF in general and feedback approaches for LM in particular. Section 3 introduces the sentence based query expansion. Section 4 describes our experimental setup and presents the results on TREC adhoc topics, while Section 5 contains a detailed analysis on the robustness of the proposed method and finally Section 6 concludes with directions for future work.

2 Related Work

Standard blind relevance feedback (BRF) techniques for IR assume that the top R documents as a whole are relevant and extract T terms to expand the original query. The various different IR models have corresponding different approaches to QE (see, for example [6,7,8]). Typically, BRF can increase performance on average for a topic set, but does not perform well on all topics. Some research even questions the usefulness of BRF in general [9]. Many approaches have been proposed to increase the overall IR performance of BRF, for example by adapting the number of feedback terms and documents per topic [10], by selecting only good feedback terms [11,12], or by increasing

---

[1] We employ the term passage in its most general sense, denoting phrases, sentences, paragraphs, and other small text units.

diversity of terms in pseudo-relevant documents by skipping feedback documents [13]. TREC experiments with BRF use conservative settings for the number of feedback terms and documents (see [7,14]) using less than 10 documents and 10-30 feedback terms to obtain the best IR effectiveness. In contrast, Buckley et al. [15] performed massive query expansion using the Vector Space Model of the SMART[2] retrieval system for ad-hoc retrieval experiments at TREC 3. The experiments involved Rocchio feedback, a linear combination of re-weighted query terms [6]. In these experiments, 300 to 530 terms and phrases were added for each topic. An improvement in retrieval effectiveness between 7% and 25% in various experiments was observed. Among the existing QE methods in LM the most commonly used ones are: a) Selecting query expansion terms by the LM score [16], b) Estimating an underlying relevance model from query terms and the top ranked documents [17]. Ponte [16] defines a LM score for the words occurring in the R top ranked documents and proposes adding the top T high scoring terms to the original query. He uses the following score:

$$s(w) = \sum_{d \in R} \log \frac{P(w|M_d)}{P(w)}$$

This score prefers the terms which are frequent in the relevant documents and infrequent in the whole collection. Xu and Croft [18] proposed Local Context Analysis (LCA) which involves decomposing the feedback documents into fixed length word windows so as to overcome the problem of choosing terms from the unrelated portions of a long document. and then ranking the terms by a scoring function which depends on the co-occurrence of a word with the query term, the co-occurrence being computed within the fixed word length windows. It also uses the Inverse Document Frequency (idf) score of a word to boost the co-occurrence score of rarely occurring terms as against frequent terms. In our method, we add document sentences which have maximal similarity with the query sentence(s) thus achieving the same effect of filtering out potentially irrelevant parts of a longer document similar to LCA. We do not compute the co-occurrences explicitly nor do we use the idf scores. Lavrenko and Croft [17] provide a solid theoretical footing on the co-occurrence based feedback as done in LCA by proposing the estimation of an underlying relevance model which is supposed to generate the relevant documents as well as the query terms. Considering the event that the query terms are samples from the unknown relevance model, co-occurrence of a word with the query terms can be used to estimate this probability. An attempt to use shorter context for BRF instead of full documents can be found in [19] where document summaries are extracted based on sentence significance scores, which are a linear combination of scores derived from significant words found by clustering, the overlap of title terms and document, sentence position, and a length normalization factor. Järvelin [20] investigated under which conditions IR based on sentence extraction is successful. He investigates user interactions for true relevance feedback. Additional BRF experiments are based on TREC 7-8 data and use the Lemur system. The best result is obtained using 5 documents and 30 terms. Our proposed method can be related to the above mentioned existing works in the following ways:

---

[2] ftp://ftp.cs.cornell.edu/pub/smart/

Table 1. Differences between QE approaches

Feature	Term based QE	SBQE
QE components	Term-based	Sentence-based
Candidate scoring	Term score/RSV	Sentence similarity
Number of terms	Few terms (5-20)	Many terms (> 100)
Extraction	Terms from feedback documents or segments	Sentences from the whole document
Working Methodology	On the whole set of feedback documents	On one document at a time
Differentiation between feedback documents	Not done	More sentences are selected from a top ranked document as compared to a lower ranked one
idf factor of terms	Used	Not used

i) It utilises the co-occurrence information of LCA and Relevance Model (RLM) in a different way. A word may co-occur with a query term in a document, but they may be placed far apart. The proximity between the two cannot be handled by these two methods. Recent work by Lv and Zhai [21] attempted to address this issue by generalizing the RLM, in a method called PRLM, where non-proximal co-occurrence is down-weighted by using propagated counts of terms using a Gaussian kernel. The difference between our work and LCA and (P)RLM is that co-occurrence of terms is not computed explicitly, since we rely on the intrinsic relationship of a document word with a query term as defined by the proximity of natural sentence boundaries.

ii) A major difference between all the existing methods and our method is that our method processes each feedback document in turn instead of considering the merits all the pseudo-relevant documents (or segments) collectively. This allows us to extract more content from the top-ranked documents and less from the lower ranked ones.

iii) Our method utilizes shorter context as explored in [19] and [20], but differs from these approaches in the sense that these methods follow the traditional term selection approach over the set of extracted shorter segments whereas we do not employ any term selection method from the shorter segments (sentences). Also we do not need to tune parameters such as the window size for passages as in [5].

Existing work on sentence retrieval considering sentences as the retrieval units instead of documents [22,23]. The difference between this and ours is that our goal is not to retrieve sentences, but on sentence selection as an intermediate step to help BRF.

Table 1 summarizes the major differences between term-based QE and SBQE.

3   Sentence Based Query Expansion (SBQE)

3.1   Motivation

In LM based IR, a document d is ranked by the estimated probability $P(Q|M_d)$ of generating a query Q from the document model M_d underlying in the document D.

M_D is modelled to choose $Q = \{t_1, t_2 \ldots t_n\}$ as a sequence of independent words as proposed by Hiemstra [8]. The estimation probability is given by Equation 1.

$$P(Q|M_d) = \prod_{i=1}^{n} \lambda_i P(t_i|M_d) + (1 - \lambda_i)P(t_i) \tag{1}$$

The term weighting equation can be derived from Equation 1 by dividing it with $(1 - \lambda_i)P(t_i)$ and taking log on both sides to convert the product to summation.

$$\log P(Q|M_d) = \sum_{i=1}^{n} \log(1 + \frac{\lambda_i}{1 - \lambda_i} \frac{P(t_i|M_d)}{P(t_i)}) \tag{2}$$

Thus if the query vector q is weighted as $q_k = tf(t_k)$ and the document vector d is weighted as $d_k = log(1 + \frac{P(t_k|M_d)}{P(t_k)} \frac{\lambda_k}{1-\lambda_k})$, the dot product $d \cdot q$ gives the likelihood of generating q from d and can be used as the similarity score to rank the documents. Adding sentences from relevant documents to the query serves the purpose of making the query look more like the relevant documents and hence increases the likelihood of generating the relevant documents by increasing the maximum likelihood estimate $P(t_i|M_d)$.

3.2   Methodology

Let R be the number of top ranked documents assumed to be relevant for a query. Each pseudo-relevant document d can be represented as a set comprising of the constituent sentences. Thus $d = \{d^1, \ldots d^{\eta(d)}\}$ where $\eta(d)$ denotes the number of sentences in d and d^is are its constituent sentences. Each such sentence d^i is represented as a vector $d^i = (d_1^i, \ldots d_{\zeta(d)}^i)$, where $\zeta(d)$ is the number of unique terms in d. The components of the vector are defined as $d_j^i = \mathtt{tf}(t_j, d^i) \; \forall j \in [1, \zeta(d)]$, where $\mathtt{tf}(t_j, d^i)$ denotes the term frequency of term t_j in sentence d^i. Also the query text is similarly mapped to the vector representation. Similarity between a sentence vector and the query vector is computed by measuring the cosine of the angle between the vectors. We choose the cosine similarity because it favours shorter texts [24]. The working steps of the proposed method are as follows:

1. Initialize i to 0.
2. For each sentence vector q^j in the query do Steps 3-5.
3. For each sentence vector $d^k \in d$ (where d is the ith pseudo-relevant document) compute the cosine similarity as $\frac{d^k \cdot q^j}{|d^k||q^j|}$ and store the document sentence similarity pair in a sorted set S ordered by decreasing similarities.
4. Add the first $m_i = \min(\lfloor \frac{1-m}{r-1}(i-1) + m \rfloor, |S|)$ sentences from the set S to the query .
5. Clear the set S. If done with all pseudo-relevant documents then exit; else increment i by 1 and goto Step 2.

The value of m_i is obtained by a linear interpolation as shown in Step 4, the slope of the interpolating line being uniquely determined from the fact that we use m number of sentences for the top ranked document and 1 sentence for the bottom ranked one. This ensures that as we go down through the ranked list we progressively become more selective in adding the sentences.

3.3   A Formal Justification

For simplicity let the initial query be $q = (q_1, \ldots q_n)$ where each unique term q_i occurs only once. From a pseudo-relevant document, we add the sentences with maximum similarity back to the original query. Since a similar sentence must have one or more query terms in it, in other words we can say that whenever one or more query terms are found in a document sentence, we add the same query terms with some more additional terms back to the original query. This methodology can be modeled as a variant of Polya's urn scheme where it is known that the distribution of balls after a sufficiently large number of draws approaches a Dirichlet distribution [25].

The initial query can be thought of as an urn of n balls, the ith ball having a colour c_i. Each pseudo-relevant document can be thought of as a bag of transparent boxes (sentences) so that it is possible to know whether a box (sentence) contains a ball (term) of colour similar to a ball from the query urn. Let us also consider another initially empty urn (expanded query) where we pour in the contents from the selected transparent boxes. A turn in the game comprises of opening a bag, looking at the boxes with matching balls and emptying its contents onto the output urn. If we find more boxes with colour c_i, we are going to end up with more balls of colour c_i in the output urn. Let us assume that after a large number of draws, we have α_i balls of colour c_i where $i = 1, 2, \ldots N$ and let the total number of balls be $\alpha_0 = \sum_i^N \alpha_i$. The expectation of finding a ball of colour c_i is

$$E[(X_i = c_i)] = \frac{\alpha_i}{\alpha_0}$$

Thus, after a sufficient number of steps in the game, we could say that the distribution of colours of balls in the output urn follows a Dirichlet distribution $Dir(\alpha_1, \ldots \alpha_N)$.

Coming back to the feedback algorithm, the colours are analogous to unique terms and the output urn is the generated expanded query $X \sim Dir(\alpha)$, $X, \alpha \in \mathbb{R}^N$, i.e. the expanded query comprises of N unique terms, the event $X_i = t_i$ denoting that the ith term is t_i. It is well known that the Dirichlet distribution $Dir(X, \alpha)$ is the conjugate prior to the multinomial distribution $Mult(X, \alpha)$. Since $Mult(X, \alpha)$ is the likelihood model of generating a query term used in the LM retrieval, the expanded query can be seen as the posterior probability of the event $X_i = t_i$ after seeing α_i occurrences of t_i in the pseudo-relevant documents. Another way of expressing this is that placing a prior distribution of $Dir(\alpha)$ on the parameters $P(X_i = t_i)$s of the multinomial distribution through the expanded query is equivalent to adding $\alpha_i - 1$ pseudo observations of term t_i in addition to the actual number of occurrences of t_i in a document (true observation) in the feedback step.

Speaking in simple terms, it is not only the presence of a query term that the feedback method utilizes, but it tries to reproduce the distribution of the original query terms and the new terms co-occurring with the original ones through evidences in the top R document texts as accurately as possible. Thus, in the feedback step this distribution of the pseudo-occurrences of new terms can benefit the conjugate prior used in the LM retrieval model.

In Section 5.3 we experimentally verify the two hypotheses that firstly it is not the mere presence of expansion terms but rather the distribution which helps in the feedback step, and secondly the greater the number of documents we examine, the better our estimation becomes. However, there is a practical limit to the number of documents that should be examined simply because for every new document examined, the chance that we are going to add a set of terms which are not already added, increases. We do not want N (the number of unique terms in the expanded query) to become too large so that we do not end up with an excessive number of hits on the inverted list.

Lavrenko and Croft state that "Many popular techniques in Information Retrieval, such as relevance feedback and automatic query expansion have a very intuitive interpretation in the traditional probabilistic models, but are conceptually difficult to integrate into the language modeling framework ..."[17]. As we have shown, SBQE is a simple, yet conducive to an intuitive interpretation in LM framework where we can argue the evidence collected from the pseudo-relevant documents generates an expanded query *looking* like the pseudo-relevant documents (formally speaking following a Dirichlet prior of pseudo-observations of the pseudo-relevant documents) thus benefiting the feedback step.

4 Experimental Results

4.1 Description and Settings

To evaluate the effectiveness of our proposed method, we used the TREC adhoc document collection (disks 1-5) and title fields of the adhoc topic sets 101-200 (TREC 2-4) and 300-450 (TREC 6-8). We do not use the TREC-5 queries as these queries comprise terms which are poor or negative indicators of relevance (see [26] for more details). Retrieval for all the TREC topic sets were done indexing the corresponding official document sets i.e. TREC 2 and 3 retrievals use disks 1 and 2, TREC 4 uses disks 2 and 3, and TREC 6-8 use disks 4 and 5.

Instead of trying to achieve optimal results by parameter tuning for each topic set, we aim to investigate the robustness and portability of SBQE for unseen topics. We used the TREC 2 query relevance judgments to train our system. The best settings obtained were then used on TREC 3 and 4 topic sets. Similarly TREC 6 query relevance judgments were used for training, and testing was done on TREC 7-8 topics. The reason behind this break-up of the training and test sets is that TREC topic sets 2,3 and 4 resemble each other in terms of the average number of relevant documents. These tasks benefit from using a higher value of R (the number of pseudo-relevant documents) in BRF experiments. Whereas the average number of relevant documents is much less for the TREC 6-8 topic sets and a smaller value of R proves to be effective for these topic sets.

Table 2 summarizes the average number of relevant documents for the individual topic sets.

We used the LM module implemented in SMART by one of the authors for indexing and retrieval. Extracted portions of documents were indexed according to Equation 1 and using single terms and a pre-defined set of phrases (using the standard SMART phrase list) according to Equation 1. The retrieval used $\lambda_i = 0.3$ for all query terms. Sentence boundaries were detected using the Morphadorner package[3]. Stopwords were removed using the standard SMART stopword list. Words were stemmed using the Porter stemmer [27].

---

[3] http://morphadorner.northwestern.edu/morphadorner/

Table 2. Summarization of our experimental setup based on the average number of relevant documents for the topic sets

Adhoc-set 1				Adhoc-set 2			
Data set	Topic #	Usage	Avg. # relevant	Data set	Topic #	Usage	Avg. # relevant
TREC-2	101-150	Training	232.9	TREC-6	301-350	Training	92.2
TREC-3	151-200	Testing	196.1	TREC-7	351-400	Testing	93.4
TREC-4	201-250	Testing	130.0	TREC-8	401-450	Testing	94.5

To compare our approach with the existing feedback approaches in LM, we selected two baselines, the first being the LM term based query expansion, hereafter referred to as LM, as in Equation 1 which was implemented in SMART. The second baseline used the RLM implementation of Indri [28] with default settings. For RLM on TREC topics set 2-4, we used the 50 top documents to estimate the relevance model as reported by Lavrenko and Croft [17]. For the TREC 6-8 topics, our experiments with Indri revealed that the best MAP obtained is by using 5 pseudo-relevant documents, and hence for these topics we used 5 documents to estimate the relevance model. As far as the number of expansion terms is concerned, best results are obtained with no expansion terms (which is also the default settings in Indri) for the RLM implementation. While it may seem that it is unfair to choose the number of expansion terms to be zero for RLM, it is important to note that RLM relies particularly on estimating a relevance model and reordering the initially retrieved documents by KL-divergence from the estimated relevance model. Additional expansion terms do not play a key role in the working principle of the model.

4.2 Feedback Effect

One of the parameters to vary for both LM and SBQE is the number of documents to be used for the BRF which we refer to as R. The other parameter to vary for SBQE is m which is the number of sentences to add. We vary both R and T (the number of terms to add for LM) in the range of [5, 50] in steps of 5.

Figures 1a and 1c suggest that for LM, the MAP degrades with an increase in the number of expansion terms, but with SBQE there is no noticeable degradation in MAP with an increase in m, as is evident from Figures 1b and 1d. Also the LM graphs are more parameter sensitive as can be seen from the larger average distances between iso-T points and greater number of intersections of the iso-R lines as compared to the SBQE graphs.

In Table 3 we report the MAPs obtained via all the three approaches for all the 300 topics, the training being done on TREC 2 and 6 topics. This table also reports the percentage changes in MAPs computed with reference to the initial retrieval MAPs for the corresponding approach. The percentage changes under the RLM column is measured relative to the Indri initial retrieval whereas the ones under LM and SBQE columns have been calculated with respect to the SMART initial retrieval.

It can be observed that SBQE outperforms both LM and RLM on these test topics. The statistically significant improvements (measured by Wilcoxon test) in MAP with

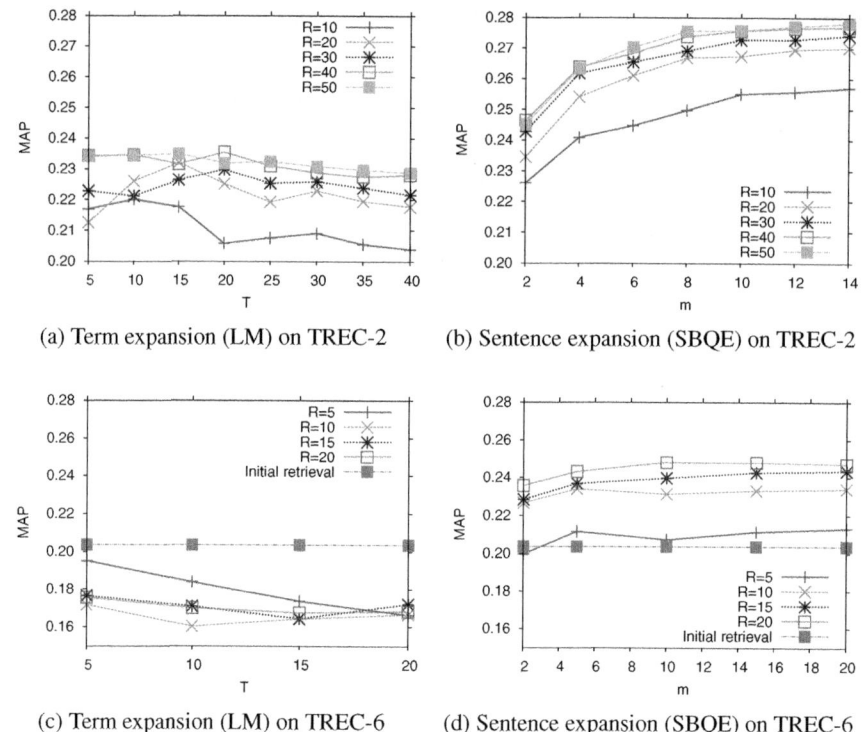

(a) Term expansion (LM) on TREC-2   (b) Sentence expansion (SBQE) on TREC-2

(c) Term expansion (LM) on TREC-6   (d) Sentence expansion (SBQE) on TREC-6

Fig. 1. Iso-R plots demonstrating the effect of varying the number of terms for LM and the parameter m for SBQE on TREC-2 and TREC-6 topics. R is the number of pseudo-relevant documents.

SBQE over LM and RLM are shown with a $^+$ and * respectively. It can also be observed that although the TREC 4 topic set uses a different collection, the same parameter settings works fairly well. This is suggestive of the relative insensitiveness of the method to precise parameter settings.

For TREC-6 (Figures 1c and 1d), we see that using LM, the best MAP we obtain is 0.1949 (which is worse than the initial retrieval) using 5 documents and 5 terms as seen in Figure 1c. Although term expansion performs very poorly on these topics, all the retrieval results being worse compared to the initial retrieval, SBQE does perform well on these topics with a significant increase in MAP compared to the initial retrieval.

The SBQE plots of Figures 1b and 1d bring out an experimental verification of the hypothesis proposed in Section 3.3, that greater the number of documents we use for predicting the Dirichlet distribution of terms in the expanded query, better the predictions for the conjugate prior become and better is the retrieval effectiveness in the feedback step. It can be observed that the MAP values of the feedback steps for increasing values of R form a strict monotonically increasing sequence.

Table 3. Three BRF approaches (LM, RLM and SBQE) on TREC 2-4 and 6-8 "title-only" topics (trained respectively on TREC 2 and TREC 6 topics)

Topic	LM Initial retrieval		MAP			Avg. # of terms	
set	SMART	Indri	LM	RLM	SBQE	LM	SBQE
101-150	0.169	0.195	0.236 (+39.4%)	0.206 (+5.5%)	**0.278**$^{+*}$ (+64.5%)	13.78	1007.26
151-200	0.215	0.234	0.288 (+34.2%)	0.242 (+3.6%)	**0.327**$^{+*}$ (+52.5%)	14.50	1141.24
201-250	0.204	0.181	0.228 (+12.2%)	0.185 (+1.9%)	**0.255**$^{+*}$ (+25.3%)	17.96	1513.66
301-350	0.207	0.217	0.195 (-6.10%)	0.226 (+4.2%)	**0.248**$^{+}$ (+19.4%)	7.48	404.84
351-400	0.161	0.185	0.163 (+0.90%)	0.187 (+0.8%)	**0.196**$^{+}$ (+21.4%)	7.42	445.90
401-450	0.241	0.241	0.213 (-11.4%)	0.245 (+1.7%)	**0.289**$^{+}$ (+12.8%)	7.38	465.88

5 Posthoc Analysis

In this section we begin with an opening subsection on query drift analysis of SBQE as compared to the other feedback methods. The following subsection aims to investigate the effectiveness of SBQE on the *hard* topics of the TREC 2004 Robust Track. This is followed by an examination of the term frequencies of the expanded query where we aim to find experimental verification of the fact that the distribution of terms in the bag-of-words model of the expanded query do play a pivotal role in the feedback step. The section ends with a subsection where we see how close this new feedback method gets to QE using true relevance feedback for TREC 6-8 ad-hoc tasks.

5.1 Query Drift Analysis

It has been found that traditional QE techniques degrade performance for many topics [9]. This can arise particularly if most of the top ranked pseudo-relevant documents are actually not relevant to the query. In these cases, QE can add terms not associated with the focus of the query, and cause the expanded query vector to draft further away from the centroid of the relevant documents and as a result the feedback retrieval can lead to worse AP for these topics.

The topics for which the initial retrieval AP is fairly good can be termed easy topics and the ones for which it is poor as difficult or hard ones. An ideal QE technique is expected to perform well over a broad range of the spectrum defined by initial retrieval AP values, ideally not degrading retrieval effectiveness for any topics. To see the effect of SBQE, we categorize all the topics (TREC 2-4 and 6-8) into classes defined by a range over the initial retrieval AP values hereafter referred to as bins. Five equal length intervals are chosen as $\{[i, i + 0.1)\}$ where $i \in \{0, 0.1 \ldots 0.4\}$. Since there are not many topics with initial retrieval AP over 0.5, the last interval is chosen as $[0.5, 1]$ so as to maintain a balance in the number of queries in each bin for meaningful comparisons. Thus the first bin contains the topics for which the initial retrieval AP is between 0 and 0.1, the second bin consists of topics for which the it is between 0.1 and 0.2 and so on. For each bin, the AP is computed by considering only the queries of that current bin. A similar analysis was presented in [29] the only difference being that they used discrete integer values of P@20 to define the bins.

Table 4. Feedback effects on the 5 topic categories for LM, RLM and SBQE. 300 topics TREC 2-4 and 6-8("title-only") were used for the analysis

LM Initial retrieval	# Queries		# Queries improved			# Queries hurt			% change in AP		
precision interval	SMART	Indri	LM	RLM	SBQE	LM	RLM	SBQE	LM	RLM	SBQE
$[0 - 0.1)$	117	120	57	51	**63**	60	69	54	+56.3	-1.6	**+75.0**
$[0.1 - 0.2)$	77	59	50	34	**57**	27	25	20	+34.1	+2.0	**+64.0**
$[0.2 - 0.3)$	37	42	23	28	**31**	14	14	6	+22.5	+7.4	**+37.1**
$[0.3 - 0.4)$	23	25	14	13	**19**	9	12	4	+7.7	+2.0	**+27.5**
$[0.4 - 0.5)$	18	21	10	**15**	15	8	6	3	+4.4	+5.8	**+23.5**
$[0.5 - 1]$	28	33	15	**24**	18	13	9	10	-5.0	**+1.3**	+1.3
Total	300	300	169	165	**203**	131	135	97			

In Table 4 we report statistics computed for each query class for the three expansion techniques. It can be observed that the SBQE achieves a positive change in AP for each query class. RLM exhibits maximal improvement (in terms of change in AP) for the group defined by the range $[0.2, 0.3)$ of initial precision values whereas both LM and SBQE work best for the topics whose initial retrieval AP is less than 0.1. LM suffers from a degradation in AP for queries with initial AP higher than 0.5, whereas SBQE and RLM improve the AP for this group, the improvement being slightly more for RLM. But improvements of AP for the other groups in SBQE are considerably higher than RLM. It is worth noting that the number of queries hurt by feedback for every query class other than the last one (where RLM is the winner with 9 hurts compared to 10 for SBQE) is the minimum for SBQE, thus making it an attractive query expansion method. The total number of queries being hurt is far less as compared to LM and RLM.

5.2   Feedback Effect on TREC-2004 Robust Track Topics

The TREC Robust track explored retrieval for a challenging set of topics from the TREC ad hoc tasks [30]. The 45 topics from TREC topics 301-450 were categorized as *hard* based on Buckley's failure analysis [31]. Buckley [32] categorized the topics into failure classes with increasing difficulty and required natural level language understanding. He suggests that retrieval could be improved for the 17 topics in categories 2-5 if systems could predict in the category to which topic a topic belongs.

We applied SBQE on the *hard* topics (categories 2-5) without the use of external resources and/or selective query expansion methods. We also ran all the three methods used in experiments on 100 new topics (601-700) designed for the TREC robust track as instances of challenging topics. Our results for individual groups of topics are shown in Table 5. From these results, we can see clearly that SBQE outperforms LM and RLM for the *hard* topics. SBQE achieves a MAP of *0.354* which ranks third among the official set of results [30] behind *pircRB04t3* [33] and *fub04Tge* [34] both of which employ web documents as an external resource for BRF. The important observation to be made here is that SBQE, without any separate training on *hard* topics, is able to achieve good precision without the use of any external resources and without employing selective query expansion (which itself consumes additional computation time).

Table 5. Feedback effect on the hard topics categorized into 4 failure classes and 100 new topics for the TREC 2004 robust track. For all queries only the title field was used. All the methods use the best parameter settings obtained by training on TREC-6 topics.

Topic Category	LM Initial retrieval		MAP		
	SMART	Indri	LM	RLM	SBQE
2: General technical failures such as stemming	0.225	0.116	0.127 (-9.7%)	0.118 (+0.2%)	**0.253** (+2.9%)
3: Systems all emphasize one aspect, miss another required term	0.081	0.083	0.162 (+8.1%)	0.088 (+0.5%)	**0.179** (+9.8%)
4: Systems all emphasize one aspect, miss another aspect	0.089	0.071	0.147 (+5.7%)	0.074 (+0.3%)	**0.152** (+6.2%)
5: Some systems emphasize one aspect, some another, need both	0.103	0.118	0.119 (+1.5%)	0.118 (+0.0%)	**0.140** (+3.7%)
Overall	0.108	0.092	0.139 (+3.1%)	0.094 (+0.2%)	**0.165** (+5.7%)
601-700 (New topics for TREC 2004 Robust track)	0.261	0.262	0.277 (+1.6%)	0.274 (+1.2%)	**0.354** (+9.3%)

We take a sample query from each category and report some terms added by SBQE, but not by LM term expansion. For topic 445 - "women clergy" belonging to category 3, true feedback adds terms like *stipend, church, priest, ordain, bishop, England* etc. The description of the topic reads "What other countries besides the United States are considering or have approved women as clergy persons". While LM expansion adds the terms *church, priest* and *ordain*, SBQE adds the additional terms (*bishop*, 7), (*England*, 10), (*stipend*, 7), (*ordain*, 11) where the numbers beside the terms indicate their occurrence frequencies in the expanded query. Common sense suggests that according to the description of this topic, *England* is indeed a good term to add. A look at topic 435 - "curbing population growth" belonging to category 4, reveals that term based LM feedback adds terms like *billion, statistics, number*, while it misses terms representing the other aspect of relevance (the aspect of contraceptive awareness in rural areas to prevent population growth - emphasized by terms like *rural, contraceptive* etc.), which are added by SBQE.

5.3   Term Frequency Analysis of Expanded Query

To justify the hypothesis of the estimated Dirichlet prior for the expanded query as the key working reason behind SBQE, we perform a series of experiments on the generated expanded query for the TREC 8 topic set. Firstly we set the term frequencies for each unique term to 1, thus reducing the expanded query to a uniform distribution where every term is equally likely to occur. Next, we seek an answer to the question of whether all terms that we added to the query are indeed useful for retrieval or could we filter out some of the rarely occurring terms from the expanded query. We therefore remove terms falling below a cut-off threshold of frequency 10, 2 and 1. Table 6a reports the observations and clearly shows that the frequencies indeed play a vital role because retrieval effectiveness decreases either when we set the term frequencies to one ignoring the evidence we collected from each feedback document or when we leave out some of the terms. Since we add a large number of terms to the original query, the expanded query at a first glance might intuitively suggest a huge query drift. But the observation which needs to be made here is that a vast majority of the terms are of low frequency. *Important* are those terms which have maximal evidence of occurrence in the feedback documents in proximity to the original query terms, the notion of proximity being defined by natural sentence boundaries. However, frequency alone is not the only criterion for the *goodness* of a term. Some low frequency terms are beneficial for the feedback step too as suggested by the fact that simply cutting off the terms based on frequency has a negative effect on precision.

5.4 Comparison with True Relevance Feedback

To see if SBQE is indeed able to add the *important* query terms to the original query we run a series of true relevance feedback (TRF) experiments which involve selecting terms by the LM term selection values as done in our standard BRF experiments, the

Terms	MAP
All terms	0.2887
$tf(t_i) \leftarrow 1$ (Frequencies set to 1)	0.1805
Terms with frequency > 1	0.280
Terms with frequency > 2	0.273
Terms with frequency > 10	0.248

(a) Term frequency variations on the expanded TREC-8 topics

Method	System	Avg. # terms	Time (s)
LM	SMART	7.38	7
RLM	Indri	2.38	209
SBQE	SMART	465.88	91

(b) Run-time measures on TREC-8 topics

Table 6. Intersection of BRF terms with the gold-standard TRF terms

Topic set	TRF		LM		SBQE	
	MAP	$|T_{TRF}|$	MAP	$|T_{TRF} \cap T_{LM}|$	MAP	$|T_{TRF} \cap T_{SBQE}|$
TREC-6	0.409	1353	0.195	316 (23.3%)	0.248	901 (66.6%)
TREC-7	0.422	1244	0.163	311 (25.0%)	0.196	933 (75.0%)
TREC-8	0.376	1234	0.213	317 (25.7%)	0.289	977 (79.1%)

only difference being that we use only the true relevant out of the top R documents of the initial ranked list for feedback.

While we do not expect that SBQE could outperform TRF, this experiment was designed with a purpose of testing how close the performance of SBQE can get to the ideal scenario. Our main aim was to define a gold-standard for the feedback terms by restricting the LM term selection value to the set of true relevant documents with the assumption that the terms hence selected for feedback provide the best possible evidence of *good* feedback terms. An overlap between the terms obtained by SBQE and the *good* terms found this way can be a measure of the effectiveness of SBQE.

We do the TRF experiments for both TREC 6-8 topic sets. The choice of true relevant documents was left on the top 20 documents from the initial retrieval ranked list. In Table 6 we report the intersection of the set of terms obtained by LM and SBQE with TRF terms. We also re-report the MAPs from Table 3 for convenience of reading. We observe from Table 6 that SBQE is able to add more *important* terms due to the higher degree of overlap with TRF terms.

5.5   Run-Time Comparisons

One may think that using more than 400 terms for retrieval can lead to poor retrieval performance. But a careful analysis reveals that the time complexity of retrieval for a query of n terms, under a sorted inverted-list implementation scheme as in SMART, is $O(\sum_{i=1}^{n} |L_i|)$, L_i being the size of the inverted list for the ith query term. On the simplified assumption that $L_i = L \ \forall i \in [1, n]$, the retrieval complexity reduces to $O(nL)$. In the worst case, if $n = O(L)$, then the run-time complexity becomes $O(L^2)$. But for SBQE, n, which is in hundreds, is still much less than the average document frequency of query terms. For example in TREC topic 301 - " International Organized Crime", the sum over the document frequencies for the terms is 215839. The SBQE expanded query comprises of 225 terms which is much less as compared to the total size of the inverted lists for the query terms. Our runtime experiments reported in Table 6b reveal that SBQE is faster than the RLM implementation of Indri.

6   Conclusions and Future Work

The main contribution of the paper is the proposal of a novel method of QE by adding sentences in contrast to the traditional approach of adding terms. The proposed method aims to make the query look more like the top ranked documents hence increasing the probability of generating the query from the top ranked documents. We also show that the method behaves like a variant of Polya's urn, and the resulting distribution of terms in the expanded query tends to the conjugate prior of the multinomial distribution used for LM retrieval. While we do not formally derive the output distribution for the variant of Polya's urn, we can explore more on this in our future research.

Although conceptually simple and easy to implement, our method significantly out-performs existing LM feedback methods on 300 TREC topics. A detailed per topic analysis reveals that SBQE increases the AP values for all types of queries when they are categorized from hardest to easiest based on initial retrieval APs. Applying SBQE on the challenging topics from the TREC robust track shows that it significantly outper-forms LM and RLM without the use of external resources or selective QE.

For term expansion, it is observed that a variable number of expansion terms chosen dynamically for the individual topics provides best effective results [10]. As future work we would like to explore whether using different m values across topics yields further improvement in the retrieval effectiveness. The method can also be extended to handle fixed length word windows (pseudo-sentences).

Whether involving any of the sentence scoring mechanisms outlined in [22,23] in our method instead of the simple cosine similarity for selecting the candidate sentences for feedback proves more beneficial will form a part of our future work as well.

Acknowledgments

This research is supported by the Science Foundation Ireland (Grant 07/CE/I1142) as part of the Centre for Next Generation Localisation (CNGL) project.

References

1. Wilkinson, R.: Effective retrieval of structured documents. In: SIGIR, pp. 311–317. Springer New York, Inc., New York (1994)
2. Tombros, A., Sanderson, M.: Advantages of query biased summaries in information retrieval. In: SIGIR 1998, pp. 2–10. ACM, New York (1998)
3. Terra, E.L., Warren, R.: Poison pills: harmful relevant documents in feedback. In: CIKM 2005, pp. 319–320. ACM, New York (2005)
4. Callan, J.P.: Passage-level evidence in document retrieval. In: SIGIR 1994, pp. 302–310. ACM/Springer (1994)
5. Allan, J.: Relevance feedback with too much data. In: SIGIR 1995, pp. 337–343. ACM Press, New York (1995)
6. Rocchio, J.J.: Relevance feedback in information retrieval. In: The SMART Retrieval System – Experiments in Automatic Document Processing. Prentice-Hall, Englewood Cliffs (1971)
7. Robertson, S.E., Walker, S., Jones, S., Hancock-Beaulieu, M.M., Gatford, M.: Okapi at TREC-3. In: Overview of the Third Text Retrieval Conference (TREC-3), pp. 109–126. NIST (1995)
8. Hiemstra, D.: Using Language Models for Information Retrieval. PhD thesis, Center of Telematics and Information Technology, AE Enschede (2000)
9. Billerbeck, B., Zobel, J.: Questioning query expansion: An examination of behaviour and parameters. In: ADC 2004, vol. 27, pp. 69–76. Australian Computer Society, Inc. (2004)
10. Ogilvie, P., Vorhees, E., Callan, J.: On the number of terms used in automatic query expansion. Information Retrieval 12(6), 666–679
11. Cao, G., Nie, J.Y., Gao, J., Robertson, S.: Selecting good expansion terms for pseudo-relevance feedback. In: SIGIR 2008, pp. 243–250. ACM, New York (2008)
12. Leveling, J., Jones, G.J.F.: Classifying and filtering blind feedback terms to improve information retrieval effectiveness. In: RIAO 2010, CID (2010)
13. Sakai, T., Manabe, T., Koyama, M.: Flexible pseudo-relevance feedback via selective sampling. ACM Transactions on Asian Language Processing 4(2), 111–135 (2005)
14. Robertson, S., Walker, S., Beaulieu, M., Willett, P.: Okapi at TREC-7: Automatic ad hoc, filtering, vlc and interactive track 21, 253–264 (1999)
15. Buckley, C., Salton, G., Allan, J., Singhal, A.: Automatic query expansion using SMART: TREC 3. In: Overview of the Third Text REtrieval Conference (TREC-3), pp. 69–80. NIST (1994)
16. Ponte, J.M.: A language modeling approach to information retrieval. PhD thesis, University of Massachusetts (1998)
17. Lavrenko, V., Croft, B.W.: Relevance based language models. In: SIGIR 2001, pp. 120–127. ACM, New York (2001)
18. Xu, J., Croft, W.B.: Query expansion using local and global document analysis. In: SIGIR 1996, pp. 4–11. ACM, New York (1996)
19. Lam-Adesina, A.M., Jones, G.J.F.: Applying summarization techniques for term selection in relevance feedback. In: SIGIR 2001, pp. 1–9. ACM, New York (2001)
20. Järvelin, K.: Interactive relevance feedback with graded relevance and sentence extraction: simulated user experiments. In: CIKM 2009, pp. 2053–2056. ACM, New York (2009)

21. Lv, Y., Zhai, C.: Positional relevance model for pseudo-relevance feedback. In: SIGIR 2010, pp. 579–586. ACM, New York (2010)
22. Murdock, V.: Aspects of Sentence Retrieval. PhD thesis, University of Massachusetts - Amherst (2006)
23. Losada, D.E.: Statistical query expansion for sentence retrieval and its effects on weak and strong queries. Inf. Retr. 13, 485–506 (2010)
24. Wilkinson, R., Zobel, J., Sacks-Davis, R.: Similarity measures for short queries. In: Fourth Text REtrieval Conference (TREC-4), pp. 277–285 (1995)
25. Blackwell, D., James, M.: Fergusson distributions via Polya urn schemes. Annals of Statistics, 353–355 (1973)
26. Xu, J., Croft, W.B.: Improving the effectiveness of informational retrieval with Local Context Analysis. ACM Transactions on Information Systems 18, 79–112 (2000)
27. Porter, M.F.: An algorithm for suffix stripping. Program 14(3), 130–137 (1980)
28. Strohman, T., Metzler, D., Turtle, H., Croft, W.B.: Indri: a language-model based search engine for complex queries. In: Online Proceedings of the International Conference on Intelligence Analysis (2005)
29. Mitra, M., Singhal, A., Buckley, C.: Improving automatic query expansion. In: SIGIR 1998, pp. 206–214. ACM, New York (1998)
30. Voorhees, E.M.: Overview of the TREC 2004 robust track. In: TREC (2004)
31. Harman, D., Buckley, C.: The NRRC Reliable Information Access (ria) workshop. In: SIGIR 2004, pp. 528–529. ACM, New York (2004)
32. Buckley, C.: Why current IR engines fail. In: SIGIR 2004, pp. 584–585. ACM, New York (2004)
33. Kwok, K.L., Grunfeld, L., Sun, H.L., Deng, P.: TREC 2004 robust track experiments using PIRCS. In: TREC (2004)
34. Amati, G., Carpineto, C., Romano, G.: Fondazione Ugo Bordoni at TREC 2004. In: TREC (2004)

Word Clouds of Multiple Search Results

Rianne Kaptein[1] and Jaap Kamps[1,2]

[1] Archives and Information Studies, University of Amsterdam, The Netherlands
[2] ISLA, Informatics Institute, University of Amsterdam, The Netherlands

Abstract. Search engine result pages (SERPs) are known as the most expensive real estate on the planet. Most queries yield millions of organic search results, yet searchers seldom look beyond the first handful of results. To make things worse, different searchers with different query intents may issue the exact same query. An alternative to showing individual web pages summarized by snippets is to represent whole group of results. In this paper we investigate if we can use word clouds to summarize groups of documents, e.g. to give a preview of the next SERP, or clusters of topically related documents. We experiment with three word cloud generation methods (full-text, query biased and anchor text based clouds) and evaluate them in a user study. Our findings are: First, biasing the cloud towards the query does not lead to test persons better distinguishing relevance and topic of the search results, but test persons prefer them because differences between the clouds are emphasized. Second, anchor text clouds are to be preferred over full-text clouds. Anchor text contains less noisy words than the full text of documents. Third, we obtain moderately positive results on the relation between the selected world clouds and the underlying search results: there is exact correspondence in 70% of the subtopic matching judgments and in 60% of the relevance assessment judgments. Our initial experiments open up new possibilities to have SERPs reflect a far larger number of results by using word clouds to summarize groups of search results.

1 Introduction

In this paper we investigate the use of word clouds to summarize multiple search results. We study how well users can identify the relevancy and the topic of search results by looking only at the word clouds. Search results can contain thousands or millions of potentially relevant documents. In the common search paradigm of today, you go through each search result one by one, using a search result snippet to determine if you want to look at a document or not. We want to explore an opportunity to summarize multiple search results which can save the users time by not having to go over every single search result. Documents are grouped by two dimensions. First of all, we summarize complete SERPs containing documents returned in response to a query. Our goal is to discover whether a summary of a SERP can be used to determine the relevancy of the search results on that page. If that is the case, such a summary can for example be placed at the bottom of a SERP so the user can determine if he wants to look at the next result page, or take another action such as rephrasing the query.

Secondly, documents are grouped by subtopic of the search request. Search results are usually documents related to the same topic, that is the topic of the search request.

A. Hanbury, A. Rauber, and A.P. de Vries (Eds.): IRFC 2011, LNCS 6653, pp. 78–93, 2011.

Query 33 : elliptical trainer

Group 1
1 : I'm looking for reviews of elliptical machines.
2 : Where can I buy a used or discounted elliptical trainer?
3 : What are the benefits of an elliptical trainer compared to other fitness machines?

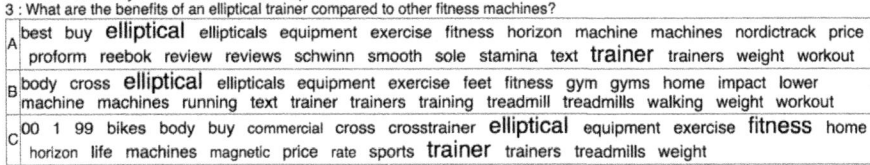

Fig. 1. Full-text clouds for the query 'Elliptical Trainer'' of the subtopic matching task

However, a query can be related to different user needs where a distinction can be made between ambiguous and faceted queries. Ambiguous queries are those that have multiple distinct interpretations, and most likely a user interested in one interpretation would not be interested in the others. Faceted queries are underspecified queries with different relevant aspects, and a user interested in one aspect may still be interested in other aspects [3]. In this paper facets and interpretations of ambiguous queries are both considered as subtopics of the query.

Clustering search results into subtopics of the query can organise the huge amount of search results. Efficiently summarising these clusters through the use of a word cloud can help the users select the right cluster for their search request. Examples of a word cloud can be found in Figure 1. These clouds are generated for subtopics of the query 'elliptical trainer'[1]. For each of the three subtopics a word cloud is generated from documents relevant to those subtopics.

Tag and word clouds are being explored for multiple functions, mainly on the social Web. Tag clouds summarize the tags assigned by users to documents, whereas word clouds can summarize documents without user assigned tags. Since there is no need for a manual effort to generate word clouds, there is a much larger potential of document sets where word clouds can be helpful. Terms in a tag cloud usually link to a collection of documents that are associated with that tag.

An advantage of word clouds is that there is no need for high quality, grammatically correct text in the documents. Using word clouds we can make summaries of web results like twitter streams, blogs, or transcribed video. Since the transcriptions usually still contain a considerable number of errors they are not suitable for snippet generation for examples. Word clouds are a good alternative, also because repeatedly occurring words have a higher chance of getting recognized [18]. Also we can make use of anchor text, which is a source of information that is used to rank search results, but which is not usually visible to the user. The anchor text representation of a web document is a collection of all the text which is used on or around the links to a document. Again, anchor text does not consist of grammatically correct sentences, but it does contain a lot of repetition, which is advantageous for the generation of word clouds.

In this paper we want to answer the following main research question:

> *How can we use word clouds to summarize multiple search results to convey the topic and relevance of these search results?*

---

[1] This is topic 33 of the 2009 TREC Web track [3].

In the context of search, we want to investigate the following issues. The snippets used in modern web search are query biased, and are proven to be better than static document summaries. We want to examine if the same is true for word clouds, hence our first research question is:

Are query biased word clouds to be preferred over static word clouds?

Besides the text on a web page, web pages can be associated with anchor text, i.e. the text on or around links on web pages linking to a web page. This anchor text is used in many search algorithms. Our second research question is:

Is anchor text a suitable source of information to generate word clouds?

The remainder of this paper is organized as follows. In the next section we discuss related work. Section 3 describes the models we use to generate the word clouds. In section 4 we evaluate the word clouds by means of a user study. Finally, in section 5 we draw our conclusions.

2   Related Work

In this section we discuss related work on snippets and alternative search result presentations, cluster labeling, keyphrase extraction and tag clouds. Many papers on search result summaries focuses on single documents, where the snippet is the most common form of single document summarization. It has been shown that query biased snippets are to be preferred over static document summaries consisting of the first few sentences of the document [17]. Query biased summaries assist users in performing relevance judgements more accurately and quickly, and they alleviate the users' need to refer to the full text of the documents.

An alternative to the traditional web search result page layout is investigated in [21]. Sentences that highly match the searcher's query and the use of implicit evidence are examined, to encourage users to interact more with the results, and to view results that occur after the first page of 10 results.

Another notable search application with an alternative search interface is PubCloud. PubCloud is an application that queries PubMed for scientific abstracts and summarizes the responses with a tag cloud interface [11]. A stopword list is used to remove common words, and a Porter stemmer is applied. Colours are used to represent recency, and font size represents frequency. Mousing over a tag displays a list of words that share the same prefix and a hyperlink links to the set of PubMed abstracts containing the tag.

Related research is done in the field of cluster labeling and the extraction of keywords from documents. Similar to our word cloud generation algorithms, these techniques extract words that describe (clusters of) documents best.

Pirolli et al. [12] present a cluster-based browsing technique for large text collections. Clusters of documents are generated using a fast clustering technique based on pairwise document similarity. Similar documents are placed into the same cluster. Recursively clustering a document collection produces a cluster hierarchy. Document clusters are summarized by topical words, the most frequently occurring words in a cluster, and typical titles, the words with the highest similarity to a centroid of the cluster. Participants in a user study were asked to rate the precision of each cluster encountered. It was shown that summarization by keywords is indeed suitable to convey the relevance of document clusters.

The goal of cluster labeling is to find the single best label for a cluster, i.e. the label equal to a manually assigned label, these algorithms generate a ranking of possible labels, and success is measured at certain cut-offs or through a Mean Reciprocal Rank. Manually assigned category labels are extracted for example from the internet directory DMOZ such as is done in [2]. The set of terms that maximizes the Jensen-Shannon Divergence distance between the cluster and the collection is considered as cluster label. Wikipedia is used as an external source from which candidate cluster labels can be extracted. Instead of the text of the documents Glover et al. [5] use the extended anchor text of web pages to generate cluster labels.

In the machine learning community, a similar task is keyphrase extraction. Here, the task is seen as a classification task, i.e. the problem is to correctly classify a phrase into the classes 'keyphrase' and 'not-keyphrase' [4]. A keyphrase can contain up to three or sometimes five words. While information retrieval approaches usually consider documents as "bags-of-words", except for term dependencies, some keyphrase extraction techniques also take into account the absolute position of words in a document. The Kea keyphrase extraction algorithm of Frank et al. [4] uses as a feature the distance of a phrase from the beginning of a document, which is calculated as the number of words that precede its first appearance, divided by the number of words in the document. The basic feature of this and the following algorithms is however a frequency measure, i.e. TF*IDF (Term Frequency*Inverse Document Frequency). Turney [19] extends the Kea algorithm by adding a coherence feature set that estimates the semantic relatedness of candidate keyphrases aiming to produce a more coherent set of keyphrases. Song et al. [15] use also a feature 'distance from first occurrence'. In addition, part of speech tags are used as features. The extracted keyphrases are used for query expansion, leading to improvements on TREC ad hoc sets and the MEDLINE dataset.

While for snippets it is clear that query biased snippets are better than static summaries, cluster labels are usually static and not query dependent. Many experiments use web pages as their document set, but the extracted labels or keyphrases are not evaluated in the context of a query which is the purpose of this study.

Most works concerning tag and word clouds address the effects of visual features such as font size, font weight, colour and word placement [1, 7, 14]. General conclusions are that font size and font weight are considered the most important visual properties. Colour draws the attention of users, but the meaning of colours is not always obvious. The position of words is important, since words in the top of the tag cloud attract more attention.

In previous work we studied the similarities and differences between language models and word clouds [10]. Word clouds generated by different approaches are evaluated by a user study and a system evaluation. Two improvements over a word cloud based on text frequency and the removal of stopwords are found. Applying a parsimonious term weighting scheme filters out not only common stopwords, but also corpus specific stopwords and boosts the probabilities of the most characteristic words. Secondly, the inclusion of bigrams into the word clouds is appreciated by our test persons. Single terms are sometimes hard to understand when they are out of context, while the meaning of bigrams stays clear even when the original context of the text is missing. In this work we take more contextual information into account, namely the query that was used to generate the search results and the anchor text of the search results.

Most tag clouds on the Web are generated using simple frequency counting techniques. While this works well for user-assigned tags, we need more sophisticated models to generate word clouds from documents. These models will be discussed in the next section.

3   Word Cloud Generation

We generate word clouds using the language modeling approach. We choose this approach because it is conceptually simple. The approach is based on the assumption that users have some sense of the frequency of words and which words distinguish documents from others in the collection [13]. As a pre-processing step we strip the HTML code from the web pages to extract the textual contents. We use three models to generate the word clouds.

3.1   Full-Text Clouds

In previous work we have shown that the parsimonious language model is a suitable model to generate word clouds [9]. The parsimonious language model [8] is an extension to the standard language model based on maximum likelihood estimation, and is created using an Expectation-Maximization algorithm. Maximum likelihood estimation is used to make an initial estimate of the probabilities of words occurring in the document.

$$P_{mle}(t_i|D) = \frac{tf(t_i, D)}{\sum_t tf(t, D)} \quad (1)$$

where D is document, and $tf(t, D)$ is the text frequency, i.e. the number of occurrences of term t in D. Subsequently, parsimonious probabilities are estimated using *Expectation-Maximisation*:

$$\text{E-step: } e_t = tf(t, D) \cdot \frac{(1 - \lambda)P(t|D)}{(1 - \lambda)P(t|D) + \lambda P(t|C)}$$
$$\text{M-step: } P_{pars}(t|D) = \frac{e_t}{\sum_t e_t}, \text{ i.e. normalize the model} \quad (2)$$

where C is the background collection model. In the initial E-step, maximum likelihood estimates are used for $P(t|D)$. Common values for the smoothing parameter λ are 0.9 or 0.99. We see that when $\lambda = 0.9$, the word clouds contain a lot of very general words and many stopwords. When $\lambda = 0.99$ the word clouds contain more informative words, and therefore in the rest of this work we set λ to 0.99. In the M-step the words that receive a probability below our threshold of 0.0001 are removed from the model. This threshold parameter determines how many words are kept in the model and does not affect the most probable words, which are used for the word clouds. In the next iteration the probabilities of the remaining words are again normalized. The iteration process stops after a fixed number of iterations.

Instead of generating word clouds for single documents, we create word clouds for sets of documents. We want to increase the scores of words which occur in multiple documents. This is incorporated in the parsimonious model as follows:

$$P_{mle}(t_i|D_1, \ldots, D_n) = \frac{\sum_{i=1}^n tf(t, D_i)}{\sum_{i=1}^n \sum_t tf(t, D_i)} \qquad (3)$$

The initial maximum likelihood estimation is now calculated over all documents in the document set $D_1, \ldots, D_n$. This estimation is similar to treating all documents as one single aggregated document. The E-step becomes:

$$e_t = \sum_{i=1}^n tf(t, D_i) * df(t, D_i, \ldots, D_n) \cdot \frac{(1-\lambda)P(t|D_1, \ldots, D_n)}{(1-\lambda)P(t|D_1, \ldots, D_n) + \lambda P(t|C)} \qquad (4)$$

In the E-step also everything is calculated over the set of documents now. Moreover, to reward words occurring in multiple documents we multiply the term frequencies tf by the document frequencies df, the number of documents in the set in which the term occurs, i.e., terms occurring in multiple documents are favoured. The M-step remains the same.

Besides single terms, multi-gram terms are suitable candidates for inclusion in word clouds. Most social websites also allow for multi-term tags. Our n-gram word clouds are generated using an extension of the bigram language model presented in [16]. We extend the model to a parsimonious version, and to consider n-grams. Our n-gram language model uses only ordered sets of terms. The model based on term frequencies then looks as follows:

$$\begin{aligned} &P_{mle}(t_j, \ldots, t_m|D_i, \ldots, D_n) \\ &= \frac{\sum_{i=1}^n tf(t_j, \ldots, t_m, D_i)}{\mathrm{argmin}_{j=1,\ldots,m} \sum_{i=1}^n tf(t_j, D_i)} * \frac{df(t_j, \ldots, t_m, D_i, \ldots, D_n)}{n} \end{aligned} \qquad (5)$$

The parsimonious version of this model takes into account the background collection to determine which n-grams distinguish the document from the background collection. To promote the inclusion of terms consisting of multiple words, in the E-step of the parsimonious model we multiply e_t by the length of the n-gram. Unfortunately, we do not have the background frequencies of all n-grams in the collection. To estimate the background probability $P(t_j, \ldots, t_m|C)$ in the parsimonious model we therefore use a linear interpolation of the smallest probability of the terms in the n-gram occurring in the document, and the term frequency of this term in the background collection.

Another factor we have to consider when creating a word cloud is overlapping terms. The word cloud as a whole should represent the words that together have the greatest possible probability mass of occurrence. That means we do not want to show single terms that are also part of a multi-gram term, unless this single term occurs with a certain probability without being part of the multi-gram term. We use the algorithm depicted in Fig. 2 to determine which words to include in the cloud. The head of a n-gram term is the term without the last word, likewise the tail is the term without the first word.

To determine the size of a term in the clouds we use a log-scale to bucket the terms into four different font sizes according to their probabilities of occurrence.

```
Create a set of n-gram terms ranked by their scores to
potentially include in the cloud
while the maximum number of terms in the cloud is not reached
do
Add the highest ranked term to the cloud
Subtract the score of the term from the score of its head and
tail
if The head or tail of the term is already in the cloud
then
Remove it from the cloud, and insert it to the set of
potential terms again
end if
end while
```

Fig. 2. Pseudo-code for constructing a n-gram cloud from a set of ranked terms

3.2 Query Biased Clouds

In the parsimonious model the background collection C is used to determine what are common words in all documents to determine what words distinguish a certain document from the background collection. In our case where documents are returned for the same search request, it is likely that these documents will be similar to each other. All of them will for example contain the query words. Since we want to emphasize the differences between the groups of search results, we should use a smaller and more focused background collection. So in addition to the background collection consisting of the complete document collection, we use a topic specific background collection. For the documents grouped by relevance, the topic specific background collection consists of the top 1,000 retrieved documents of a search topic. For the documents grouped by subtopic of the search request, the topic-specific background collection consists of all documents retrieved for any subtopic. Using background models on different levels of generality helps to exclude non-informative words.

We estimate a mixed model with parsimonious probabilities of a word given two background collections as follows:

$$\text{E-step: } e_t = tf(t, D) \cdot \frac{(1 - \lambda - \mu)P(t|D)}{(1 - \lambda - \mu)P(t|D) + \lambda P(t|C_1) + \mu P(t|C_2)}$$

$$\text{M-step: } P_{pars}(t|D) = \frac{e_t}{\sum_t e_t}, \text{ i.e. normalize the model} \tag{6}$$

There are two background models: C_1 and C_2. C_1 is the model based on the complete corpus. C_2 is the topic specific model. The weight of the background models is determined by two parameters, λ and μ. We keep the total weight of the background models equal at 0.99, so we choose for λ and μ a value of 0.495.

Our standard model uses the full text of documents to generate word clouds. In addition to using a focused background collection, we focus on the text around the query terms to generate query biased clouds. The surrogate documents used to generate query biased clouds contain only terms that occur around the query words. In our experiments all terms within a proximity of 15 terms to any of the query terms is included.

Example query 1 : dog heat
Description : What is the effect of excessive heat on dogs?

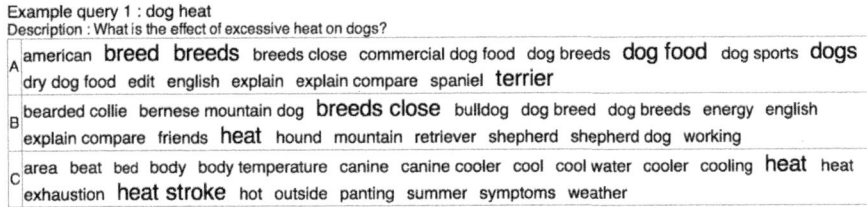

| | american **breed** **breeds** breeds close commercial dog food dog breeds **dog food** dog sports **dogs** dry dog food edit english explain explain compare spaniel **terrier** |
|A| |

| | bearded collie bernese mountain dog **breeds close** bulldog dog breed dog breeds energy english explain compare friends **heat** hound mountain retriever shepherd shepherd dog working |
|B| |

| | area beat bed body body temperature canine canine cooler cool cool water cooler cooling **heat** heat exhaustion **heat stroke** hot outside panting summer symptoms weather |
|C| |

Fig. 3. Query biased clouds for the query 'Dog Heat' of the relevance assessment task

Query 17 : poker tournaments
Group 1
1 : I want to find information on the World Series of Poker.
2 : I want to find Texas Hold-Em tournaments.
3 : Find books on tournament poker playing.

| | bellagio cup **colorado poker tournaments** **kansas city poker tournaments** online poker tournaments poker blog poker tournament **tournaments** **upcoming poker tournaments** **wendover poker tournaments** |
|A| |

| | arnold **books** fast formula online **online poker** patience factor play players **poker** poker onlinecasinoswiss com **poker tournament strategy and..** **poker tournaments** skill strategy tournament tournaments |
|B| |

| | 1978 wsop 1979 wsop **1980** **1981** 1988 1995 1999 wsop **2004 wsop** **2006** **world series of poker circuit event** |
|C| |

Fig. 4. Anchor text clouds for the query 'Poker tournaments' of the subtopic matching task

3.3 Anchor Text Clouds

So far, we used the document text to generate word clouds. On the web however, there is another important source of information that can be used to summarize documents: the anchor text. When people link to a page, usually there is some informative text contained in the link and the text around the link.

The distribution of anchor text terms greatly differs from the distribution of full-text terms. Some webpages do not have any anchor text, while others have large amounts of (repetitive) anchor text. As a consequence we can not use the same language models to model full text and anchor text. Anchor texts are usually short and coherent. We therefore treat each incoming anchor text as one term, no matter how many words it contains. For each document, we only keep the most frequently occurring anchor text term. The terms are cut off at a length of 35, which only affects a small number of terms. Maximum likelihood estimation is used to estimate the probability of an anchor text term occurring, dividing the number of occurrences of the anchor text by the total number of anchor text terms in the document set. When after adding all the anchor text terms to the word cloud the maximum number of terms in the cloud is not reached, the anchor text cloud is supplemented with the highest ranked terms from the document's full text.

4 Experiments

We conduct a user study to evaluate our word cloud generation models. After describing the set-up, results are given and analyzed.

4.1   Experimental Set-Up

To evaluate the quality of our word clouds we perform a user study consisting of two tasks. The set-up of the user study is as follows. Participants are recruited by e-mail. The user study is performed online and starts with an explanation of the task, including some examples and a training task. A short pre-experiment questionnaire follows, before the experiment starts with the subtopic matching task, which consists of 10 queries. Three versions of the study are generated, which together cover 30 queries for each part of the study. A version is randomly assigned when a test person starts the study.

For each query two groups of clouds have to be matched to particular subtopics. The three methods described in the previous section are used to generate the groups of word clouds: Full-Text (FT), Query biased (QB), and Anchor text (AN). The two groups of clouds are generated using two out of the three word cloud generation methods, which are selected using a rotation scheme. The test persons do not know which word cloud generation methods are used. Besides the matching task, the test persons also assign a preference for one of the two groups. The second part of the study is the relevance assessment task, which consists of 10 queries with two groups of clouds. Again for each query two out of the three word cloud generation methods are selected using a rotation scheme. Finally, a post-experiment questionnaire finishes the user study.

We use different sets of queries for each pair of word cloud generation methods allowing for pairwise comparison. Since the query effect is large due to differences in the quality of retrieved documents, we cannot compare all three methods on the same grounds.

Task 1: Subtopic Matching: When queries are ambiguous or multi faceted, can the word cloud be used to identify the clusters? To evaluate the disambiguation potential of word clouds we let test persons perform a matching task. Given a query, and a number of subtopics of this query, test persons have to match the subtopics to the corresponding word clouds. An example topic for this task can be found in Figure 1.

Topics are created as follows. We use topics from the diversity task in the TREC 2009 Web track [3]. Topics for the diversity task were created from the logs of a commercial search engine. Given a target query, groups of related queries using co-clicks and other information were extracted and analysed to identify clusters of queries that highlight different aspects and interpretations of the target query. Each cluster represents a subtopic, and the clusters of related queries are manually processed into a natural language description of the subtopic, which is shown to our test persons.

The clouds in the user study are generated as follows. The relevance of documents to subtopics is judged by assessors hired by TREC. From the relevance assessments we extract relevant documents for each subtopic. A subtopic is only included if there are at least three relevant documents. Furthermore, we set a minimum of two subtopics per query topic, and a maximum of four. If there are more than four subtopics with at least three relevant documents, we randomly select four subtopics. The methods used to generate the word clouds from the selected documents are described in the previous section.

Task 2: Relevance Assessment: How well can test persons predict if results are relevant by looking at a word cloud? To evaluate this task we let test persons grade word clouds which represent a complete search result page for a particular query. These word clouds are graded by the test persons in our user study on a three-point scale (Relevant, Some relevance, Non relevant). An example topic for this task can be found in Figure 3. Three word clouds are created for each topic using 20 documents, i.e one cloud generated using only relevant documents, one cloud generated where half of the documents are relevant, and the other half are non-relevant, and one cloud generated using only non-relevant documents). In the ideal case the test person evaluates the cloud created from only relevant documents as "Relevant", the cloud created from non-relevant documents as "Non relevant", and the cloud created from the mix of relevant and non-relevant documents as "Some relevance".

The topics we use are taken from the ad hoc task of the TREC 2009 Web track. We use the relevance assessments of the track to identify relevant documents, and the documents from the bottom of the ranking of a standard language model run returning 1,000 results as non-relevant documents. To ensure there are differences between the relevant and the non-relevant documents, we take the documents from the bottom of the ranking of a standard language model run returning 1,000 results as non-relevant documents. There is a small chance that there are still some relevant documents in there, but most documents will not be relevant, although they will contain at least the query words.

4.2 Experimental Results

We evaluate our word cloud generation methods through the user study. This leads to the following results.

Demographics. In total 21 test persons finished the complete user study. The age of the test persons ranges from 25 to 42 year, with an average age of 30. Most test persons were Dutch, but overall 11 nationalities participated. All test persons have a good command of the english language. A large part of the test persons is studying or working within the field of information retrieval or computer science. The familiarity with tag clouds is high, on average 3.8 measured on a Likert-scale, where 1 stands for 'totally unfamiliar' and 5 stands for 'very familiar'. On average the test persons spent 38 minutes on the user study in total. The first task of subtopic matching took longer with an average of 19 minutes, while the second task of relevance assessments went a bit quicker with an average of 14 minutes. Since the tasks are always conducted in the same order, this could be a learning effect.

Query Biased Word Clouds. We take a look at the results of both tasks in the user study (subtopic matching and relevance judgments) to answer our first research question: 'Are query biased word clouds to be preferred over static word clouds?'. The first task in the user study was to match subtopics of the search request to the word clouds. Our test persons perform the subtopic matching significantly better using the full-text model (significance measured by a 2-tailed sign-test at significance level 0.05). The full-text clouds judgments match the ground truth in 67% of all assignments, the query biased clouds match in 58% of the cases.

Table 1. Percentage of correct assignments on the relevance assessments task

Model	Relevant	Half	Non Relevant	All
FT	0.42	0.36	0.44	0.40
QB	0.42⁻	0.39⁻	0.50⁻	0.44⁻

Table 2. Confusion matrix of assignments on the relevance assessments task for the FT model

	Assessed as		
Generated from	Relevant	Half	Non Relevant
Relevant	178	180	72
Half	222	154	54
Non Relevant	66	174	186

In the second task of the user study the test persons assess the relevance of the presented word clouds on a three-point scale. Although each group of clouds contains one cloud of each relevance level, the test persons can choose to assign the same relevance level to multiple word clouds. Since in the subtopic matching task each subtopic should be matched to only one cloud, there could be a learning effect that the test persons assign each relevance level also to only one cloud. We show the results of this task in Table 1. On the relevance assessment task the query biased model performs better than the full-text model, but the difference is not significant.

The results split according to relevance level are shown in the confusion matrices in Tables 2 and 3. We see that the clouds containing some relevance (half) match the ground truth the least often. The non relevant clouds are recognized with the highest accuracy, especially in the query biased model. When we look at the distribution of the relevance levels, it is not the case that most assignments are to 'Non relevant'. For both models the distinction between clouds generated from relevant documents, and clouds generated from a mix of relevant and non-relevant documents is the hardest to make for our test persons.

Anchor Text Clouds. We now examine our second research question 'Is anchor text a suitable source of information to generate word clouds?'. On the subtopic matching task, the anchor text model performs slightly better than the full-text model on the subtopic task, with an accuracy of 72% versus an accuracy of 68% of the full text model.

Results of the relevance assessment task can be found in Table 4. The anchor text model performs best, with almost 60% of the assignments correctly made. Again the clouds with some relevance are the hardest to recognize. The confusion matrices of both models show a pattern similar to the confusion matrices in Figure 2 and 3, and are therefore omitted here.

The inter-rater agreement for both tasks measured with Kendall's tau lies around 0.4, which means there is quite some disagreement. Besides comparing the word cloud generation methods on their percentages of correct assignments, we can also compare the word cloud generation methods from the test person's point of view. For each query, the test persons assess two groups of word clouds without knowing which word cloud

Table 3. Confusion matrix of assignments on the relevance assessments task for the QB model

	Assessed as		
Generated from	Relevant	Half	Non Relevant
Relevant	*180*	168	84
Half	222	*168*	42
Non Relevant	78	138	*216*

Table 4. Percentage of correct assignments on the relevance assessments task

Model	Relevant	Half	Non Relevant	All
FT	0.61	0.47	0.56	0.54
AN	0.62 ⁻	0.50 ⁻	0.63 ⁻	0.59 ⁻

generation method was used, and they selected a preference for one of the clouds. The totals of all these pairwise preferences are shown in Table 5. The full-text model performs worst on both tasks. On the subtopic task, the query biased model outperforms the anchor text model, but the difference is not significant.

Analysis. To analyze our results and to get some ideas for improving the word clouds we look at the comments of test persons. First thing to be noticed is that test persons pay a lot of attention to the size of the terms in the cloud, and they focus a lot on the bigger words in the cloud. The algorithm we use to determine the font sizes of the terms in the clouds can be improved. Our simple bucketing method works well for log-like probability distributions, but some of the word cloud generation methods like the anchor-text model generate more normal probability distributions. For these distributions, almost all terms will fall into the same bucket, and therefore have the same font size.

One of the most frequently reported problems with the clouds that they contain too much noise, i.e words unrelated to the query. The tolerance of noise differs greatly among the test persons. We can identify three types of noise:

- HTML code. For some queries test persons comment on the occurrence of HTML code in the clouds. This noise can easily be removed by improving the HTML stripping procedure. Since this problem occurs at the document pre-processing step, it affects all word cloud generation methods to the same degree.
- Terms from menus and advertisements. Not all the textual contents of a web page deals with the topic of the web page. Although frequently occurring terms like "Home" or "Search" will be filtered out by our term weighting schemes, sometimes terms from menus or advertisements are included in the clouds. This problem can be solved by applying a content extractor for web pages to extract only the actual topical content of a page such as described in [6]. This procedure can also take care of the HTML stripping. Improving the document pre-processing step will increase the overall quality of all word clouds.
- Non informative terms. Some terms occur frequently in the documents, but do not have any meaning when they are taken out of context, such as numbers (except years). It may be better to not include numbers below 100 and terms consisting of one character at all in word clouds.

Table 5. Pairwise preferences of test person over word cloud generation models

Model 1	Model 2	# Preferences Subtopic			# Preferences Relevance		
		Model 1	Model 2	Sign test	Model 1	Model 2	Sign test
AN	FT	**47**	21	99%	**43**	23	95%
AN	QB	39	**47**		34	34	
FT	QB	29	**41**		23	**43**	95%

This may explain in part why the anchor text clouds work well, that is it has less problems with noise. Anchor text is more focused and cleaner than the full text of a web page.

The second frequently reported problem is that clouds are too similar. During the creation of the user study we already found that clouds created from judged relevant, and judged non relevant documents were very similar. We noticed that the documents judged as non-relevant were very similar in their language use to the relevant documents, so using the judged non-relevant documents led to only minor differences in the language models of the relevant documents and the non-relevant documents. We suspect most search systems that contributed to the pool of documents to be judged are heavily based on the textual contents of the documents, whereas a commercial search engine uses many other factors to decides on the ranking of pages, leading to documents whose textual content will be more dissimilar.

A similar observation is made in the recent work of Venetis et al. [20].They define a formal framework for reasoning about tag clouds, and introduce metrics such as coverage, cohesiveness and relevance to quantify the properties of tag clouds. An 'ideal user satisfaction model' is used to compare tag clouds on the mostly uncorrelated evaluation metrics. A user study is conducted to evaluate the user model. Although the model often predicts the preferred tag cloud when users reach agreement, average user agreement is low. They observe in many cases users do not have a clear preference among clouds, it is therefore important for user studies involving word or tag clouds to make sure there are clear differences between the clouds.

For for some of the queries in our study the clouds are indeed very similar to each other with a large overlap of the terms in the cloud. The query biased clouds emphasise the differences between the clusters of documents, and generate the most dissimilar clouds. This is most probably the reason why the test persons prefer the query biased clouds. Unfortunately, the query bias in the clouds does comes with a loss of overall quality of the clouds and does not lead to a better representation of the topic and the relevance in the clouds.

Summarising the results, anchor text is a good source of information to generate word clouds and although query biased clouds are preferred by the test persons, they do not help to convey the topic and relevance of a group of search results.

5   Conclusions

In this paper we investigated whether word clouds can be used to summarize multiple search results to convey the topic and relevance of these search results. We generate word clouds using a parsimonious language model that incorporates n-gram terms, and experiment with using anchor text as an information source and biasing the clouds towards the query.

The snippets used in modern web search are query biased, and are proven to be better than static document summaries. We want to examine if the same is true for word clouds, hence our first research question is: *Are query biased word clouds to be preferred over static word clouds?* Surprisingly, we have not found any positive effects on the performance of test persons by biasing the word clouds towards the query topic. The test persons however did appreciate this model in their explicit preferences, because it emphasizes the differences between the clusters of documents.

Secondly, we studied the use of anchor text as a document surrogate to answer the question: *Is anchor text a suitable source of information to generate word clouds?* We find a positive answer to this research question; anchor text is indeed a suitable source of information. The clouds generated by the documents' anchor text contain few noisy terms, perform better than the full-text model, and the anchor text clouds are preferred by the test persons as well.

Finally, the main research question of this paper was: *How can we use word clouds to summarize multiple search results to convey the topic and relevance of these search results?.* We have studied a new application of word clouds, and tested how well the user perception of such a cloud reflects the underlying result documents, either in terms of subtopics or in terms of the amount of relevance. Although tag and word clouds are pervasive on the Web, no such study exists in the literature. The outcome of our study is mixed. We achieve moderately positive results on the correspondence between the selected word clouds and the underlying pages. Word clouds to assess the relevance of a complete SERP achieve an accuracy of around 60% of the assignments being correct, while subtopics are matched with an accuracy of around 70%. It is clear however that interpreting word clouds is not so easy. This may be due in part to the unfamiliarity of our test persons with this task, but also due to the need to distinguish between small differences in presence of noise and salient words. Especially the word clouds based on varying degrees of relevant information seem remarkably robust. This can also be regarded as a feature: it allows for detecting even a relatively low fraction of relevant results.

In future work we would like to compare the use of word clouds for summarization of search results to other summarization methods, such as snippets. While for the subtopic matching task a snippet from a single document could be sufficient, for the relevance assessment task we would need to experiment with generating snippets from multiple documents, since the relevance level of a complete result page cannot be judged by a snippet from a single result. We also would like to apply content extraction techniques to extract the actual content from the web pages and thereby reduce the noise occurring in the clouds.

Acknowledgements. This research was supported by the Netherlands Organization for Scientific Research (NWO, under project # 612.066.513). In case you are wondering: the correct assignments of the clouds in Figures 1, 3 and 4 respectively are: 1-A, 2-C, 3-B; A-Non Rel., B-Some Rel., C-Rel.; and 1-C, 2-A, 3-B.

References

[1] Bateman, S., Gutwin, C., Nacenta, M.: Seeing things in the clouds: the effect of visual features on tag cloud selections. In: HT 2008: Proceedings of the Nineteenth ACM Conference on Hypertext and Hypermedia, Pittsburgh, PA, USA, pp. 193–202. ACM, New York (2008)

[2] Carmel, D., Roitman, H., Zwerdling, N.: Enhancing cluster labeling using wikipedia. In: Proceedings of SIGIR 2009, pp. 139–146. ACM, New York (2009)

[3] Clarke, C.L.A., Craswell, N., Soboroff, I.: Overview of the trec 2009 web track. In: Proceedings of the Eighteenth Text REtrieval Conference, TREC 2009 (2010)

[4] Frank, E., Paynter, G.W., Witten, I.H., Gutwin, C., Nevill-Manning, C.G.: Domain-specific keyphrase extraction. In: IJCAI 1999: Proceedings of the Sixteenth International Joint Conference on Artificial Intelligence, pp. 668–673 (1999)

[5] Glover, E., Pennock, D.M., Lawrence, S., Krovetz, R.: Inferring hierarchical descriptions. In: Proceedings of CIKM 2002, pp. 507–514. ACM, New York (2002)

[6] Gupta, S., Kaiser, G., Neistadt, D., Grimm, P.: Dom-based content extraction of html documents. In: Proceedings of the 12th International Conference on World Wide Web, WWW 2003, pp. 207–214. ACM, New York (2003)

[7] Halvey, M.J., Keane, M.T.: An assessment of tag presentation techniques. In: WWW 2007: Proceedings of the 16th International Conference on World Wide Web, pp. 1313–1314. ACM, New York (2007)

[8] Hiemstra, D., Robertson, S., Zaragoza, H.: Parsimonious language models for information retrieval. In: Proceedings of the 27th Annual International ACM SIGIR Conference on Research and Development in Information Retrieval, pp. 178–185. ACM Press, New York (2004)

[9] Kaptein, R., Hiemstra, D., Kamps, J.: How different are language models and Word clouds? In: Gurrin, C., He, Y., Kazai, G., Kruschwitz, U., Little, S., Roelleke, T., Rüger, S., van Rijsbergen, K. (eds.) ECIR 2010. LNCS, vol. 5993, pp. 556–568. Springer, Heidelberg (2010)

[10] Kaptein, R., Serdyukov, P., Kamps, J., de Vries, A.P.: Entity ranking using Wikipedia as a pivot. In: Proceedings of the 19th ACM Conference on Information and Knowledge Management (CIKM 2010), pp. 69–78. ACM Press, New York (2010)

[11] Kuo, B.Y.-L., Hentrich, T., Good, B.M., Wilkinson, M.D.: Tag clouds for summarizing web search results. In: WWW 2007: Proceedings of the 16th International Conference on World Wide Web, pp. 1203–1204. ACM, New York (2007)

[12] Pirolli, P., Schank, P., Hearst, M., Diehl, C.: Scatter/gather browsing communicates the topic structure of a very large text collection. In: Proceedings of the SIGCHI Conference on Human Factors in Computing Systems: Common Ground, CHI 1996, Vancouver, British Columbia, Canada, pp. 213–220. ACM, New York (1996)

[13] Ponte, J.M., Croft, W.B.: A language modeling approach to information retrieval. In: Proceedings of the 21st ACM Conference on Research and Development in Information Retrieval, pp. 275–281 (1998)

[14] Rivadeneira, A.W., Gruen, D.M., Muller, M.J., Millen, D.R.: Getting our head in the clouds: toward evaluation studies of tagclouds. In: CHI 2007: Proceedings of the SIGCHI Conference on Human Factors in Computing Systems, San Jose, California, USA, pp. 995–998. ACM, New York (2007)

[15] Song, M., Song, I. Y., Allen, R. B., Obradovic, Z.: Keyphrase extraction-based query expansion in digital libraries. In: JCDL 2006: Proceedings of the 6th ACM/IEEE-CS Joint Conference on Digital Libraries, pp. 202–209 (2006)

[16] Srikanth, M., Srihari, R.: Biterm language models for document retrieval. In: Proceedings of SIGIR 2002, pp. 425–426. ACM, New York (2002)

[17] Tombros, A., Sanderson, M.: Advantages of query biased summaries in information retrieval. In: Proceedings of SIGIR 1998, pp. 2–10. ACM, New York (1998)

[18] Tsagkias, M., Larson, M., de Rijke, M.: Term clouds as surrogates for user generated speech. In: Proceedings of SIGIR 2008, pp. 773–774. ACM, New York (2008)

[19] Turney, P.: Coherent keyphrase extraction via web mining. In: IJCAI 2003, Proceedings of the Eighteenth International Joint Conference on Artificial Intelligence, pp. 434–442 (2003)

[20] Venetis, P., Koutrika, G., Garcia-Molina, H.: On the selection of tags for tag clouds. In: Proceedings of the Fourth ACM International Conference on Web Search and Data Mining, WSDM 2011, Hong Kong, China, pp. 835–844. ACM, New York (2011)

[21] White, R.W., Ruthven, I., Jose, J.M.: Finding relevant documents using top ranking sentences: an evaluation of two alternative schemes. In: Proceedings of SIGIR 2002, pp. 57–64. ACM, New York (2002)

Free-Text Search over Complex Web Forms

Kien Tjin-Kam-Jet, Dolf Trieschnigg, and Djoerd Hiemstra

University of Twente, Enschede, The Netherlands
{tjinkamj,trieschn,hiemstra}@cs.utwente.nl

Abstract. This paper investigates the problem of using free-text queries as an alternative means for searching 'behind' web forms. We introduce a novel specification language for specifying free-text interfaces, and report the results of a user study where we evaluated our prototype in a travel planner scenario. Our results show that users prefer this free-text interface over the original web form and that they are about 9% faster on average at completing their search tasks.

Keywords: query processing, free-text interfaces, query translation.

1 Introduction

The internet contains a large amount of information that is only accessible through complex web forms. Journey planners, real estate websites, online auction and shopping websites, and other websites commonly require the user to fill out a form consisting of a number of fields in a graphical interface. The user should first interpret the form and then translate his information need to the appropriate fields. Filling out these forms can be slow because they require mixed interaction with both the mouse and keyboard. A natural language interface (NLI) alleviates these problems by allowing the user to enter his information need in a single textual statement. Rather than navigating between and entering information in the components of the web form, the user can focus on formulating his information need in an intuitive way. NLIs require or assume syntactically well-formed sentences as input, in essence restricting the range of textual input. However, describing all possible natural language statements and dealing with query ambiguity can be a time-consuming process [1,2,3,4]. Therefore, we introduce a free-text interface (FTI) which allows the user to *freely input text* without any restrictions. In this paper, we describe and evaluate a prototype system for specifying FTIs to access information behind complex web forms. The system has been designed to specify flexible FTIs with relatively little effort. This work is a stepping stone for further investigation of a single textual interface to access the deep web [5]. Ideally, we wish to use these techniques to build a distributed search system which can search multiple resources, including information behind complex web forms, simultaneously. The contributions of this paper are as follows: *i*) we introduce a specification language for describing free-text interfaces (FTIs) to complex web forms; *ii*) as a proof of concept, we show that this language can be effectively used to describe a flexible FTI to a travel planner web

A. Hanbury, A. Rauber, and A.P. de Vries (Eds.): IRFC 2011, LNCS 6653, pp. 94–107, 2011.

form; and *iii*) we demonstrate that users can search faster with an FTI than with a complex web form, and that they prefer the FTI over the complex web form. The remainder of this paper is structured as follows: Section 2 describes the requirements and our prototype framework. Our experiment setup and the results are described in Sect. 3. Sections 4 and 5 discuss our work and overview related work. Finally, Sect. 6 concludes our work.

2   A Free-Text Interface to Web Forms

2.1   Requirements

The goal of our FTI is to offer simple textual access to content behind a web form that consists of multiple input fields and options. Given a user's query as free-text input, the FTI should display a ranked list of plausible interpretations (i.e. ways to fill out the complex web form) as results. Kaufmann and Bernstein showed the importance of guiding the user during the query formulation process [2]. Therefore, the FTI should offer query suggestions as a means to guide users in formulating their queries. Lastly, it should be easy for developers to specify the capabilities of the FTI. Examples of a complex web form for a travel planning website and an FTI to the same form, are given in Figs. 1 and 2, respectively.

Fig. 1. A complex web form that offers interactive query suggestions, based on the Dutch Railways site

Fig. 2. Trajectvinder ('Route Finder'): an FTI that offers interactive query suggestions, tailored to the complex web form

2.2   Framework

On a high level, our FTI involves three processes: the query interpretation process, the query suggestion process, and the result generation process.

Basic query interpretation. Let us first define the following: A *region* is a contiguous segment or sequence of characters of the user's input. A *pattern* expresses relevant information (as prefix cues, types, and postfix cues; these are discussed later). An *annotation* is a label assigned to a region, denoting that the region contains some (part of a) pattern. Finally, an *interpretation* is a set of annotated regions. The query interpretation process consists of five steps:

1. *Scanning for input regions:* the user input is scanned for known patterns from left to right, on a word-by-word basis (a word is delineated by a white space). At each word, all matching patterns starting from that word are stored. Pattern matching is greedy, meaning a pattern will match as much of the query as possible. This process yields a set of possibly overlapping input *regions*, where each region could have several annotations. (A region could be matched by multiple patterns, e.g. a region containing the token '2000' could be annotated as: a year, an amount of money, or a car model);

2. *Generating non-overlapping region sets:* the set Γ, which contains sets of non-overlapping regions with maximal coverage of the input, is generated;

3. *Generating interpretations:* for each region set $\gamma \in \Gamma$ all combinations of annotations are generated. This yields the set of possible interpretations;

4. *Filtering interpretations:* first, interpretations are 'cleaned' by removing extraneous annotations. Examples of extraneous annotations are prefix annotations that precede annotations which are not specified by the pattern to which prefix corresponds, and annotations that exceed the number of times they are allowed to appear in the underlying web form. Second, interpretations that are completely subsumed by other interpretations are removed;

5. and,*Ranking interpretations:* the interpretations are first ranked by the number of annotations they contain, then by the order in which the patterns are specified in the configuration file.

Steps one to four are illustrated in Fig. 3. Say, we are given the query "Wycombe to shopping paradise Bicester North Camp". Further, we have two simple patterns p_1 (*from* station), and p_2 (*to* station). Here, *from* and *to* are optional

1)   Wycombe to shopping paradise Bicester North Camp
        $\overline{r_1(p_1,p_2)}$ $\overline{r_2(pr_2)}$ $\overline{r_3(p_1,p_2)}$
                                                                        $\overline{r_4(p_1,p_2)}$

2) $\Gamma = \{ \; \gamma_1 = \{r_1, r_2, r_3\} \; , \; \gamma_2 = \{r_1, r_2, r_4\} \; \}$

3) $\gamma_1 \mapsto i_1 = \{p_1, pr_2, p_1\}$ $\gamma_2 \mapsto i_5 = \{p_1, pr_2, p_1\}$
              $i_2 = \{p_1, pr_2, p_2\}$ $i_6 = \{p_1, pr_2, p_2\}$
              $i_3 = \{p_2, pr_2, p_1\}$ $i_7 = \{p_2, pr_2, p_1\}$
              $i_4 = \{p_2, pr_2, p_2\}$ $i_8 = \{p_2, pr_2, p_2\}$

4-i) $i_1 = \{p_1, \cancel{pr_2}, \cancel{p_1}\}$ $i_5 = \{p_1, \cancel{pr_2}, \cancel{p_1}\}$
        $i_2 = \{p_1, pr_2, p_2\}$ $i_6 = \{p_1, pr_2, p_2\}$
        $i_3 = \{p_2, \cancel{pr_2}, p_1\}$ $i_7 = \{p_2, \cancel{pr_2}, p_1\}$
        $i_4 = \{p_2, \cancel{pr_2}, \cancel{p_2}\}$ $i_8 = \{p_2, \cancel{pr_2}, \cancel{p_2}\}$

4-ii) $\cancel{i_1 = \{p_1\}}$ $\cancel{i_5 = \{p_1\}}$
        $i_2 = \{p_1, pr_2, p_2\}$ $i_6 = \{p_1, pr_2, p_2\}$
        $\cancel{i_3 = \{p_2, p_1\}}$ $\cancel{i_7 = \{p_2, p_1\}}$
        $\cancel{i_4 = \{p_2\}}$ $\cancel{i_8 = \{p_2\}}$

Fig. 3. An illustration of the basic interpretation process

prefix tokens, and `station` is a type which denotes a set of tokens. In this example, the only valid tokens of the type `station` are 'Wycombe', 'Bicester North', and 'North Camp'. Lastly, each pattern may occur once or not at all.

Step 1 underlines the regions of the input containing known tokens. The first region r_1 is matched by both patterns p_1 and p_2, and is annotated as such. Region r_2 contains the prefix of pattern p_2, annotated as pr_2. The overlapping regions r_3 and r_4 are split in step 2, yielding the non-overlapping sets of regions γ_1 and γ_2. For each set $\gamma \in \Gamma$, step 3 generates all possible annotation-combinations. Such a combination is in fact an interpretation, thus step 3 yields the interpretations $i_1 \ldots i_8$. Step 4-i removes the erroneous annotations in each interpretation, in this case, it removes the prefix pr_2 when it is not followed by the pattern p_2, and it removes an annotation if it already occurred. In step step 4-ii, an entire interpretation is removed if it is a subset of any other interpretation. At the end of step four, we are left with two interpretations i_2 and i_6, denoting "from Wycombe to Bicester North", and "from Wycombe to North Camp", respectively.

Generating suggestions. The suggestion process is an extension of the basic interpretation process, and generates three types of query suggestions: token expansions, pattern expansions, and relation expansions. The suggestions are interactive query expansions [6]; they are generated based on the last region, and filtered based on the entire interpretation. When the last region denotes a prefix of known tokens, all applicable token expansions are shown. When the last region denotes a complete prefix of a pattern, this triggers token suggestions of the expected type, only if the type was defined by a list of tokens. If the expected type was defined by a regular expression, no suggestions can be shown. When the last region denotes the body of some pattern, and if the set of postfix strings of this pattern is non-empty, then the default (longest) postfix is shown. Finally, when the last region contains a token (like a car brand) for which there are related tokens (like the corresponding car models), then those tokens are shown.

Generating results. The result generation process is also an extension of the basic interpretation process; each interpretation is post-processed as follows. First, the default values for all fields that were not specified by the user are added to the interpretation. For instance, by default the current time could be added to the example query given earlier. Second, the interpretation is discarded if it does not satisfy all constraints in the FTI's configuration. Third and finally, a result snippet is generated according to the FTI's configured rules (described in the next sections).

2.3 Configurable Items

Our FTI can be configured by specifying the following items: *i*) the web form's lexicon; *ii*) the constraints, *iii*) the patterns; and *iv*) the result generation rules.

Lexicon. The input fields of a web form often syntactically constrain user input, e.g. limiting the number of characters, or only accepting input from a pre-defined list of tokens. Regular expressions are used to pose even more syntactic restrictions, such as, allowing only numbers or zip-codes. Input fields (e.g. drop-down menus) may map external strings to internal form values. The lexicon contains known values (both internal and external): it consists of the regular expressions, the list of tokens, and the mapping from external to internal values.

Constraints. The constraints denote value restrictions and relations. Example restrictions are mandatory values (i.e. values that must be found in the input), or value comparisons (e.g. a price value should be greater than zero). Value relations are apparent in some web forms. For example, in the car-sales domain, each value from the class "car brands" can have an associated set of values from the class "car models". Whenever a brand is selected, the list of available models changes accordingly. These relations are useful for: limiting the set of valid queries, ranking query interpretations, and generating query suggestions.

Patterns. Consider the input query "find me a trip to Amsterdam from Paris". Here, the values are 'Amsterdam' and 'Paris', and the contexts are 'to' and 'from', respectively. A system that responds with "from Amsterdam, to Paris" in the first place and "from Paris, to Amsterdam" in the second place, is not user friendly as it generated a false positive and it may have wasted the user's valuable time. Simply scanning user input for known values, without considering the context of the extracted values, may lead to unnecessary query interpretations.

Therefore, we adopt a bottom-up approach for capturing the context: a set of patterns must be specified, one for each input field. Each pattern consists of three parts: a prefix, a body, and a postfix part. The affixes (the prefix and postfix) each denote a finite list of strings, including the empty string. Note that the affixes can be used to incorporate idioms of natural language. If a particular value is found (i.e. it matched the body part) as well as a corresponding (non-empty) affix, that value is then bound to the field to which this pattern belongs.

Two patterns can be combined into a range pattern. A range pattern is useful for disambiguation. For example, the input '1000 - 2000 euros' would be interpreted as 'minimal 1000 euros and maximal 2000 euros'. Without range patterns, we would find '1000' (it could be a car model, a min price, or a max price) and '2000 euros' (it could be min price or max price), which would have to be further processed in order to remove erroneous interpretations.

Result generation rules. A query interpretation is displayed as a *result snippet* (see Fig. 4), containing a title, a description, and a URL. To generate the title and the description, an ordered set of *field templates* must be specified. A field template specifies how a field's value should be displayed. To generate the URL of the snippet, the web form's action-URL, the http-request method

Trajectvinder

amsterdam utrecht   [search]

Routes from Amsterdam to Utrecht
Details: travelling on 14-9-2010, departure at 15:25
http://www.example.com

Routes from Utrecht to Amsterdam
Details: travelling on 14-9-2010, departure at 15:25
http://www.example.com

Fig. 4. Interpretation as result snippets

(i.e. get or post), and all the form's input fields must be specified. In the next section, we show how these configurable items fit together.

2.4   An Example Configuration

Figure 5 depicts an example configuration file and shows how the lexicon, the constraints, the patterns, and the result generation rules, are specified. The `tokens` element contains token instances. Each `instance` belongs to a specific *type*, has one internal value and a list of external values (treated as synonyms by the system). Multiple instances can belong to a single type. The `pattern` element's `id` attribute contains the name of the input field to which the captured value should be assigned. A `capture` element specifies the *type* to be captured. The `prefix` and `postfix` elements specify a finite list of strings. This list may be specified by fully writing out all possibilities, or by a Kleene star-free regular expression, which will be automatically expanded to the list of possible strings. A pattern's `option` element relates a particular `prefix` with a particular `postfix`. The use of options is portrayed in the following example. Consider an input field to enter some minimum mileage, and three prefix-capture-postfix

```
<?xml version='1.0' encoding='UTF−8' standalone='yes' ?>
<root>
  <tokens>
    <instance type='station' internal='1'>
      <external>amsterdam amstel</external>
      <external>amstel</external>
    </instance>
    <instance type='station' internal='2'>
      ...
  </tokens>
  <patterns>
    <pattern id='fromloc'>
      <option>
        <prefix>((depart(ing|ure)? )?from)?</prefix>
        <capture>station</capture>
      </option>
    </pattern>
    <pattern id='toloc'>
      ...
  </patterns>
  <constraints>
    <mandatory_fields>
      <fieldset>
        <field>fromloc</field>
        <field>toloc</field>
      </fieldset>
    </mandatory_fields>
    <field_field_not_equal='fromloc' to='toloc' />
  </constraints>
  <results>
    <url method='get'>http://www.example.com/search.html?loc1={fromloc}&amp;...</url>
    <title max='3' starttext='Example.com: search results for '>
      <fieldtemplate id='fromloc' prefix='from ' postfix=' ' />
      <fieldtemplate id='toloc' prefix='to ' postfix=' ' />
      ...
    </title>
    <defaults>
      <field id='arrivalTime' external='arriving on ' internal='true'/>
    </defaults>
  </results>
</root>
```

Fig. 5. An example configuration file

combinations: "minimum ... kilometers", "minimum number of kilometers ...", and "minimum number of kilometers ... kilometers". The latter of these combinations is peculiar and it would be parsed if we specified just one pattern option consisting of: "<prefix>minimum(number of kilometers)?</prefix>" and "<postfix>kilometers</postfix>". Moreover, the system would also generate the postfix suggestion "kilometers" if it parsed "minimum number of kilometers ...". To prevent this behavior, we could specify two options, one containing "<prefix>minimum</prefix>" and "<postfix>kilometers</postfix>", and one containing just the prefix "<prefix>minimum number of kilometers</prefix>". The constraints element may contain: *i*) a list of mandatory field combinations; or *ii*) a list of comparisons, e.g. comparing the value of one field to the value of another or to some constant value. An interpretation is valid if it satisfies at least one of the mandatory field combinations, and all field comparisons. Lastly, the results element specifies what the interpretation's title, description, and URL should look like. Here, a developer could also specify default (internal) values (with corresponding external values) for input fields. The url element specifies both the action-URL and http-request method. The title element (just like the omitted description element) contains an ordered list of field templates. Each template corresponds to exactly one of the form's input fields, indicated by the id attribute. The title (as well as the description) is generated by listing the contents of its start text attribute and concatenating the contents of the active field templates, up to the specified max number of templates. A template is active if the value of the input field it refers to is not empty.

3   Experiment and Results

We developed a prototype framework and evaluated it for an existing travel-planner web form. The web form is as depicted in Fig. 1, the resulting FTI is depicted in Fig. 2. A total of six information items can be specified in the form: a departure location, an arrival location, an optional via location, the time, the date, and a flag indicating whether the date and time are for arrival or departure.

In this experiment, we tried to answer the following questions: *i*) do people prefer to use such an FTI over the existing complex web form in the first place? *ii*) is searching by means of an FTI faster than searching by means of a complex web form? *iii*) how much variation exists in the query formulations? *iv*) are people consistent in their query formulations? *v*) what are the most positive and negative aspects of the FTI? and *vi*) why is the FTI better, or worse, than the complex web form?

3.1   Experimental Setup

Experimental procedure. The experiment consisted of an offline part, an online part, and a questionnaire. The information needs were picked at random from a list of test needs. They were shown either graphically as a route on a map; or textually as a random sequence of (two or three) station names, a date, and a time. Dates were either relative, such as "next week Wednesday", or absolute, such as

"1-2-2011". Times were described either alphabetically, such as "half past ten", or numerically, such as "17.30". During the offline part, the subjects first provided background information (e.g. age, study). Then, they wrote down their 'most recent travel question' if they could remember it. Next, an information need was shown as a route on a map, along with a desired date and time. The subjects were asked to fill out the complex web form on paper based on this information need. Likewise, but based on a different information need, they filled out the FTI on paper. Finally, the subjects were shown a filled out complex web form, and they reformulated that into a question suitable for the FTI. We aimed to collect query formulations with as little bias to the question as possible. That is why we asked the subjects to formulate a query from memory, and to formulate a query based on graphical instead of textual descriptions of the information need. During the online part, the task was to search for specific routes. Each route was described textually, with a different order of the information items (i.e. the date, time, and locations), and with different wordings (e.g. *ten past one*, or *13:10*). The subjects first familiarized themselves with the complex interface of the existing travel planner site. Then, they searched for 5 specific train routes and wrote down the departure and arrival times, while we recorded the total time to find all routes. Next, the subjects familiarized themselves with the FTI. After that, they searched for 5 specific routes and wrote down the departure and arrival times, and we recorded the total search time. Regarding the questionnaire, all questions were answered on a five-point Likert scale, except for the open questions and explanatory questions. The subjects indicated whether they thought the FTI was easy to use, if they could find results faster using the FTI, and whether the results of the FTI were correct. They indicated whether or not the FTI was nicer and better, and explained why they thought so. There were two open questions, asking the subjects to indicate the most negative and the most positive aspects of the system. Finally, they indicated which system they preferred.

Analysis. We tested whether the task completion times of the FTI differed significantly ($p < 0.05$) from those of the complex web form, using the Paired Samples T-Test [7]. We also tested whether the five-point Likert scale values differed significantly from neutral (i.e. the number '3'), also using the T-Test. Further, we evaluated the query formulation consistency by looking at the order of the information items. Each item was first replaced by a symbol as follows. We replaced the 'from' (location) with A, 'to' with B, 'via' with V, the 'date' with D, and the 'time' with T. For example, the input "from Amsterdam via Haarlem to The Hague, tomorrow at 10am." was represented as AVBDT. We then measured the rank correlation between the subject's query formulation and the task description using Kendall's τ [8]. Lastly, for each subject, we measured the average Kendall's τ over the combinations of that subject's own formulations.

3.2   Results

The subjects. A total of 17 subjects (11 male, 6 female) participated in the study. The age distribution ranged from 21 to 66 (median: 27, mean: 32); most

Table 1. Average results of the question-naire, comparing the FTI to the complex web form on a five point Likert-Scale from 1 (full agreement) to 5 (full disagreement). Results in bold are significant ($p < 0.05$).

Question	Score
Faster	**2.4**
Nicer	**1.8**
Better	2.5
Preferred	**2.0**

Table 2. Average time in minutes to complete all five search tasks for each interface. The results differ significantly ($p = 0.032$).

Interface	Average time in minutes
Free-text interface	6.7
Complex web form	7.3

subjects were between the age of 21 and 33. The background of the subjects ranged from (under)graduate students in various studies to people working in healthcare, consultancy, and IT-software development. Participation (including the questionnaire) took around 30 minutes on average for each subject.

The questionnaire. Comparing the free-text interface (FTI) against the complex web form, the subjects indicated on a five-point Likert scale whether the FTI was: *faster, nicer, better,* and *preferred.* The results are given in Table 1, where '1' indicates full agreement, and '5' denotes the opposite. All results differed significantly ($p < 0.05$) from neutral, except for the third aspect. On average, the subjects felt they could search a little faster using the FTI than using the complex web form. This was supported by the times measured for the web form and the FTI, shown in Table 2. The subjects were significantly ($p = 0.032$) faster, by about 9%, when using the FTI instead of the complex form.

Speed and success rate. We counted the number of incorrect routes reported by the subjects. Out of the 170 answers, 14 were wrong: 6 errors were made using the FTI, and 8 using the complex form. The most likely explanation for the errors is that the subjects misread the task, and entered a wrong time or station name. Out of the 17 subjects, 10 subjects made zero errors, 2 subjects made one error, 3 subjects made two errors, and 2 subjects made three errors. However, since we did not measure the time per query, we cannot omit the times for the failed queries for comparing the two systems. Yet if we would use only the data from the 10 subjects who made no errors, there would still be a 9% difference in time, in favor of the FTI.

Pros and cons. The subjects listed the most negative and most positive aspects of the FTI. The following **negative** aspects were mentioned: 24% of the subjects indicated that there was no example or short manual (forcing the subjects to 'just type in something, and it worked'); 18% indicated that the interface was too simple, e.g. it lacked pictures; and 12% disliked that they had to explicitly click a result snippet to view the travel plan, even when only a single result snippet was returned. The following **positive** aspects were mentioned: 41% of the subjects liked how the system 'understood' dates like tomorrow and Tuesday, and written time like 'ten past nine'; 41% liked that you only had to type (without clicking on menus); 35% mentioned the query-suggestions as a useful feature; and 18% appreciated the fact that the input order of information items (e.g. time, date, places) did not matter.

Consistency. When considering only the order of the information items[1] in a query, there were 17 different query formulations. As can be seen in Fig. 6, the three most frequent online query formulations were: ABDT 41%, ABVDT 15%, and, tied at third place with 6%, were ABTD, DTABV, and TABVD.

Now we inspect whether or not the subjects formulated their queries with the same order of information items as that of the online task descriptions. The mean Kendall's τ between the online task descriptions and the query formulations was **0.42**. The task with the highest average τ (0.96) was sequenced ABDT, the other four tasks were BADT (0.67), TABVD (0.39), DTABV (0.09), and TBAD (-0.02). Two subjects always followed the same information order of the task descriptions and had an average τ of 1.0 (though they used different wordings). Three subjects had an average τ between 0.6 and 1.0, and the remaining twelve subjects had an average less than or equal to 0.3.

The mean Kendall's τ for the (within subjects) online query formulations was **0.64**. Six subjects always formulated their questions in the same order, regardless of the task description, and had an average τ of 1; six subjects averaged between 0.7 and 0.9; and, five subjects had an average τ less than 0.2.

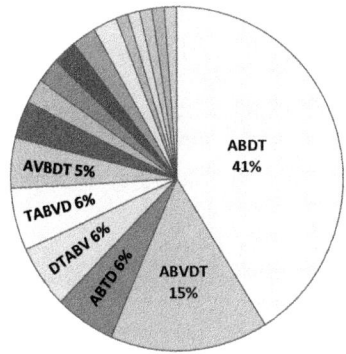

Fig. 6. Distribution of the most frequent online query formulations

Overall, the subjects were highly consistent in their query formulations individually; however, there was considerable query variation between subjects. Further, the task descriptions had little effect on the subjects' query formulations; the moderate correlation (0.42) is most probably an artifact caused by subjects consistently formulating their queries as ABDT. This explains the high correlations between the query formulations and the two tasks ABDT and BADT.

4   Discussion

4.1   Methodology and Results

Query variation. We tried to prevent the subjects from mindless copying of the task descriptions by presenting the tasks on paper instead of on screen. Nevertheless, the large number of different query formulations we collected was surprising, since: *i*) the subjects could have just retyped the task descriptions; *ii*) there were only 17 subjects; and *iii*) the travel-planner web form was relatively

---

[1] i.e. the 'date' (D), 'time' (T), and the 'from' (A), 'to' (B), and 'via' (V) locations.

simple. With so much query variation in this limited scenario (in both the order of information items and wordings used), even higher variation might be expected in a more complex scenario.

Time difference. The paper&pencil-approach demanded manual time measurement. We measured the total time to complete all 5 search tasks, as it would be more difficult to obtain accurate measurements for individual tasks. Consequently, we could not determine whether the time per task decreased or not. Even though we noticed several times that subjects were clearly experimenting with the free-text interface during the tests (as they were talking out loud, saying 'what if I typed...'), the average time of the FTI is still significantly lower than that of the complex web form.

4.2   Specialized Features

In some cases, it could be handy to invoke a suitable function with the detected values as arguments. For instance, to extract the actual 'dd-mm-yyyy' time format from the input 'next week Friday', some function similar to 'getDate()' should be called to obtain the current date in order to calculate the intended date. The framework contains several pre-defined functions (e.g. for extracting dates and times) which can be invoked simply by specifying the field(s) that accept(s) a function value. Future versions of the framework will allow developers to add new functions.

4.3   Practicality of the Framework

For users. Our work could add to the solution of the deep web problem. Given a free-text search query, we can generate "real-time deep web results" (i.e. result snippets with valid query-URLs as data-entry points). Deep web search ultimately enhances our search experience by allowing users to search more content and to specify attribute or facet restrictions, besides merely a list of key words. Our work may benefit other search environments as well, particularly when there is some sort of semi-structured search involved, examples could be desktop search and intranet search.

For providers. We believe that web companies will be encouraged to create their own configuration files for the following reasons: *i*) we showed that end users prefer such an interface; *ii*) (we claim that) it is easy to write a configuration file; and *iii*) visibility and user-friendliness are crucial for web companies.

Evidence for the last point can be found in a study by Alba et al. [9], where they observed that: (1) the revenues of websites depend on the data that users see on the site's web pages; (2) websites are extremely motivated to ensure correctness, accuracy, and consistency on the web pages shown to the end user; and (3) websites do not accord the same level of significance to the data delivered by the APIs. Alba et al. show that web companies care greatly for their 'public image', since: *i*) selling products or services is difficult if users do not know about you; and *ii*) online users are more inclined to make a purchase if they feel positive about the website. Furthermore, the large number of articles on the web about search engine optimization strongly indicates that web companies make serious investments to increase their visibility to the users.

Our free-text interface for searching over web forms has the potential to both increase the visibility of a web site (i.e. deep web search, enabling search over otherwise uncrawlable data) and to provide more user-friendly search interfaces for websites that implement this interface.

5 Related Work

A similar problem of filling out a web form for a given text query was tackled by Meng [10]. Meng used various statistical disambiguation techniques. However, a drawback of his statistical approach is that it requires (training) data that is often difficult to obtain (e.g. it requires domain-specific queries in order to obtain the 'global' statistics, and the data must often be annotated manually). Instead of statistical disambiguation, we scan for valid pattern combinations and present a ranked list of alternative interpretations to the user.

Weizenbaum described Eliza [11], one of the earliest systems with a natural language interface. The input text is parsed using decomposition rules triggered by keywords. Responses are generated based on reassembly rules pertaining to the decomposition rule. These rules are stored in a script which can be easily modified. During a session, a re-triggered decomposition rule may generate a different response. Unlike Weizenbaum, we generate responses depending on a set of detected patterns instead of a single decomposition rule, and we do not vary the responses. In the context of keyword-based retrieval systems over structured data, one of the earliest systems was DataSpot [12]. The DataSpot system used free-form queries and navigations to explore a *hyperbase* (a graph of associated elements) for publishing content of a database on the Web. Recent systems [13,14,15,16,17] generate a ranked list of structured queries or query interpretations, such that the user can select the right interpretation. However, most model the query as a bag of terms, disregarding the context of the extracted values, whereas we use patterns to capture the context. Further, they use a probabilistic or heuristic approach to rank the interpretations. Other grammar-based natural language interfaces have been developed [18,19,20,21]; however, the majority of these systems were application-specific which made it difficult to port the systems to different applications [22]. The difficulty of porting a system from one application (domain) to another is also apparent in information extraction systems, i.e. systems that extract all entities from large bodies of texts. To overcome the difficulty of porting, Appelt and Onyshkevych [4] propose the Common Pattern Specification Language (CPSL). At the heart of the CPSL grammar are the *rules*. Each rule has a priority, a pattern and an action. Input matched by the pattern part can be operated on by the action part of the rule. Ambiguity arises when multiple rules match at a given input location, and is resolved as follows: the rule that matches the largest part of the input is preferred, and if two rules match the same portion of the input, the rule with the highest priority is preferred. In case of equal priorities of matching rules, the rule declared earlier in the specification file is preferred. Like Appelt and Onyshkevych, we propose a pattern specification language, and the patterns are used to scan the input text. However, we generate interactive query suggestions and we produce a ranked list of interpretations instead of a single interpretation.

6 Conclusion and Future Work

We introduced a novel specification language for describing a free-text interface (FTI) to complex web forms. Our system uses patterns to scan the user input and extract 'bags of key/value-pairs'. The system is capable of both generating query suggestions on the fly and generating ranked query interpretations.

We carried out a user study to compare the FTI with an existing travel planner web form. Our results showed that the subjects were significantly faster at finding information when using the FTI instead of the complex form by about 9%. Furthermore, they preferred the FTI over the complex web form. The results also showed that the subjects were highly consistent in their individual query formulations, and that there was considerable query variation between subjects, even in such a relatively simple scenario.

In future work we will investigate whether configuring the FTI is really simple or not, by building FTIs in different domains and analyzing the builders' opinions about the configuration process. Also, we will investigate optimizations of the parsing process, and examine different ways to combine patterns of multiple websites in one domain.

Acknowledgment

This research was supported by the Netherlands Organization for Scientific Research, NWO, grants 639.022.809 and 612.066.513.

References

1. Sun, J., Bai, X., Li, Z., Che, H., Liu, H.: Towards a wrapper-driven ontology-based framework for knowledge extraction. In: Zhang, Z., Siekmann, J.H. (eds.) KSEM 2007. LNCS (LNAI), vol. 4798, pp. 230–242. Springer, Heidelberg (2007)
2. Kaufmann, E., Bernstein, A.: Evaluating the usability of natural language query languages and interfaces to semantic web knowledge bases. In: Web Semantics: Science, Services and Agents on the World Wide Web (2010)
3. Papakonstantinou, Y., Gupta, A., Garcia-Molina, H., Ullman, J.D.: A query translation scheme for rapid implementation of wrappers. In: Ling, T.-W., Vieille, L., Mendelzon, A.O. (eds.) DOOD 1995. LNCS, vol. 1013, pp. 161–186. Springer, Heidelberg (1995)
4. Appelt, D.E., Onyshkevych, B.: The common pattern specification language. In: Proceedings of a Workshop on Held at Baltimore, Maryland, Morristown, NJ, USA, pp. 23–30. Association for Computational Linguistics (1996)
5. Madhavan, J., Ko, D., Kot, L., Ganapathy, V., Rasmussen, A., Halevy, A.: Google's deep web crawl. In: Proc. VLDB Endow., vol. 1(2), pp. 1241–1252 (2008)

6. White, R.W., Marchionini, G.: Examining the effectiveness of real-time query expansion. Information Processing and Management 43(3), 685–704 (2007)
7. Kutner, M.H., Nachtsheim, C.J., Neter, J., Li, W.: Applied linear statistical models, 5th edn. McGraw-Hill, New York (2005)
8. Kendall, M.: Rank Correlation Methods, 4th edn. Second impression. Charles Griffin (1975)
9. Alba, A., Bhagwan, V., Grandison, T.: Accessing the deep web: when good ideas go bad. In: OOPSLA Companion 2008: Companion to the 23rd ACM SIGPLAN Conference on Object-Oriented Programming Systems Languages and Applications, pp. 815–818. ACM, New York (2008)
10. Meng, F.: A natural language interface for information retrieval from forms on the world wide web. In: ICIS, Atlanta, GA, USA, pp. 540–545. Association for Information Systems (1999)
11. Weizenbaum, J.: Eliza—a computer program for the study of natural language communication between man and machine. Commun. ACM 9(1), 36–45 (1966)
12. Dar, S., Entin, G., Geva, S., Palmon, E.: Dtl's dataspot: Database exploration using plain language. In: Proceedings of the 24th International Conference on Very Large Data Bases, VLDB 1998, pp. 645–649. Morgan Kaufmann Publishers Inc, San Francisco (1998)
13. Demidova, E., Fankhauser, P., Zhou, X., Nejdl, W.: Divq: diversification for keyword search over structured databases. In: SIGIR 2010, pp. 331–338. ACM, New York (2010)
14. Tran, T., Cimiano, P., Rudolph, S., Studer, R.: Ontology-based interpretation of keywords for semantic search. In: Aberer, K., Choi, K.-S., Noy, N., Allemang, D., Lee, K.-I., Nixon, L.J.B., Golbeck, J., Mika, P., Maynard, D., Mizoguchi, R., Schreiber, G., Cudré-Mauroux, P. (eds.) ASWC 2007 and ISWC 2007. LNCS, vol. 4825, pp. 523–536. Springer, Heidelberg (2007)
15. Zhou, Q., Wang, C., Xiong, M., Wang, H., Yu, Y.: Spark: adapting keyword query to semantic search. In: Aberer, K., Choi, K.-S., Noy, N., Allemang, D., Lee, K.-I., Nixon, L.J.B., Golbeck, J., Mika, P., Maynard, D., Mizoguchi, R., Schreiber, G., Cudré-Mauroux, P. (eds.) ASWC 2007 and ISWC 2007. LNCS, vol. 4825, pp. 694–707. Springer, Heidelberg (2007)
16. Tata, S., Lohman, G.M.: Sqak: doing more with keywords. In: SIGMOD 2008, pp. 889–902. ACM, New York (2008)
17. Kandogan, E., Krishnamurthy, R., Raghavan, S., Vaithyanathan, S., Zhu, H.: Avatar semantic search: a database approach to information retrieval. In: SIGMOD 2006, pp. 790–792. ACM, New York (2006)
18. Burton, R.R.: Semantic grammar: An engineering technique for constructing natural language understanding systems. Technical report, Bolt, Beranek and Newman, Inc., Cambridge, MA (December 1976)
19. Hendrix, G.G., Sacerdoti, E.D., Sagalowicz, D., Slocum, J.: Developing a natural language interface to complex data. ACM TODS 3(2), 105–147 (1978)
20. Carbonell, J.G., Boggs, W.M., Mauldin, M.L., Anick, P.G.: The XCALIBUR project: a natural language interface to expert systems. In: IJCAI 1983, pp. 653–656. Morgan Kaufmann Publishers Inc., San Francisco (1983)
21. Carbonell, J.G., Hayes, P.J.: Dynamic strategy selection in flexible parsing. In: Proceedings of the 19th Annual Meeting on ACL, Morristown, NJ, USA, pp. 143–147. Association for Computational Linguistics (1981)
22. Androutsopoulos, I., Ritchie, G.D., Thanisch, P.: Natural language interfaces to databases – an introduction. Natural Language Engineering 1(01), 29–81 (1995)

Multilingual Document Clustering Using Wikipedia as External Knowledge

Kiran Kumar N., Santosh G.S.K., and Vasudeva Varma

International Institute of Information Technology, Hyderabad, India
{kirankumar.n,santosh.gsk}@research.iiit.ac.in, vv@iiit.ac.in

Abstract. This paper presents Multilingual Document Clustering (MDC) on comparable corpora. Wikipedia has evolved to be a major structured multilingual knowledge base. It has been highly exploited in many monolingual clustering approaches and also in comparing multilingual corpora. But there is no prior work which studied the impact of Wikipedia on MDC. Here, we have studied availing Wikipedia in enhancing MDC performance. We have leveraged Wikipedia knowledge structure (such as cross-lingual links, category, outlinks, Infobox information, etc.) to enrich the document representation for clustering multilingual documents. We have implemented Bisecting k-means clustering algorithm and experiments are conducted on a standard dataset provided by FIRE[1] for their 2010 Ad-hoc Cross-Lingual document retrieval task on Indian languages. We have considered English and Hindi datasets for our experiments. By avoiding language-specific tools, our approach provides a general framework which can be easily extendable to other languages. The system was evaluated using F-score and Purity measures and the results obtained were encouraging.

Keywords: Multilingual Document Clustering, Wikipedia, Document Representation.

1 Introduction

The amount of information available on the web is steeply increasing with globalization and rapid development of internet technology. Due to various contributors across the world, this information is present in various languages. Hence there is a need to develop applications to manage this massive amount of varied information. MDC has been shown to be very useful in processing and managing multilingual information present on the web. It involves dividing a set of n documents written in different languages, into various clusters so that the documents that are semantically more related belong to the same cluster. It has got applications in various streams such as Cross-Lingual Information Retrieval (CLIR) [1], training of parameters in statistical machine translation, or the alignment of parallel and non parallel corpora, among others.

---

[1] Forum for Information Retrieval Evaluation - http://www.isical.ac.in/∼clia/

A. Hanbury, A. Rauber, and A.P. de Vries (Eds.): IRFC 2011, LNCS 6653, pp. 108–117, 2011.
© Springer-Verlag Berlin Heidelberg 2011

In traditional text clustering methods, documents are represented as "bag of words" (BOW) without considering the semantic information of each document. For instance, if two documents use different collection of keywords to represent the same topic, they may be falsely assigned to different clusters. This problem arises due to lack of shared keywords, although the keywords they use are probably synonyms or semantically associated in other forms. The most common way to solve this problem is to enrich the document representation with an external knowledge or an ontology. Wikipedia is one such free, web-based, collaborative, multilingual encyclopedia. In this paper, we have conducted an in-depth study on different ways of exploiting its huge multilingual knowledge in enhancing the performance of MDC. Using Wikipedia over other knowledge resources or ontologies has got advantages like:

1. Wikipedia has the multilingual content along with its metadata at a comparable level. By availing the cross-lingual links this information can be aligned.
2. Wikipedia supports 257 active language editions. As Wikipedia acts as a conceptual interlingua with its cross lingual links, our approach is scalable to other languages with relative ease.
3. With its wide access to many editors, any first story or hot topic gets updated in Wikipedia which can enhance the clustering performance of news documents. Hence, this approach also addresses the future growth of multilingual information.

The rest of the paper is organized as follows: Section 2 talks about the related work. Section 3 describes our clustering approach in detail. Section 4 presents the experiments that support our approach. Finally we conclude our paper and present the future work in Section 5.

2   Related Work

MDC is normally applied on parallel [2] or comparable corpus [3,4,5]. In the case of the comparable corpora, the documents usually are news articles. Work has been done where translation techniques are employed to compare multilingual documents for MDC. Two different strategies are followed for converting a document from one language to another language. First one is translating the whole document to an anchor language and the second one is translating certain features of the document, that best describes it, to an anchor language. The work proposed in [6] uses bilingual dictionaries for translating Japanese and Russian documents to English (anchor language). Translating an entire document into an anchor language is often not preferred due to the time overhead involved. Considering the second strategy, where the solution involves translating only some features, first it is necessary to select these features (usually verbs, nouns, etc.) and then translate them using a bilingual dictionary or by consulting a parallel corpus.

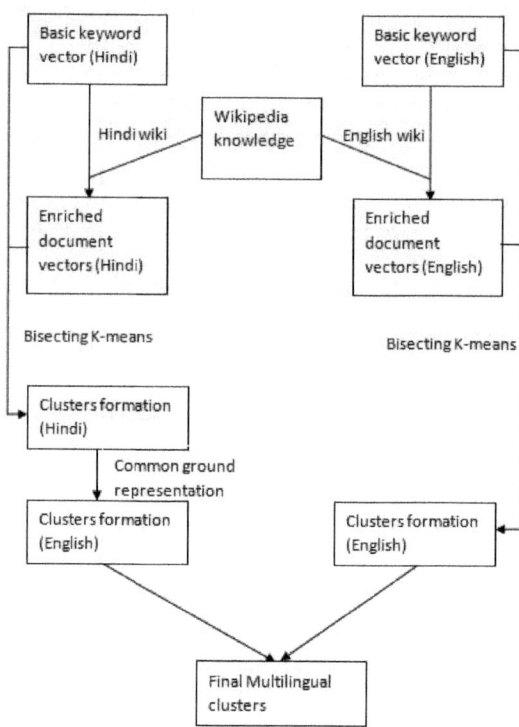

Fig. 1. MDC Approach

There is limited prior work where external knowledge was used to enhance the MDC performance. Work proposed in [7] used an existing knowledge structure, i.e. multilingual thesaurus EUROVOC to improve the multilingual document similarity measurement. The EUROVOC has support only for European languages. Steinberger *et al.* [8] proposed a method to extract language-independent text features using gazetteers and regular expressions besides thesaurus and classification systems. However, the gazetteers support only a limited set of languages. Such resources don't satisfy the need to deal with information in diverse languages. Hu *et al.* [9], Wang and Domeniconi [10] utilized the category information present in Wikipedia knowledge base for monolingual text clustering and classification respectively. But none of the revised works used Wikipedia knowledge to enhance the performance of MDC. We explain our MDC approach which is based on efficient exploitation of Wikipedia in Section 3.

3   Proposed Approach

In this section, we detail the various phases involved in the proposed approach for clustering multilingual documents. The pool of English and Hindi text documents are first represented as Keyword vectors. Additional vectors like Category vector, Outlink vector and Infobox vector are obtained by enriching the document representation (Keyword vector) with Wikipedia knowledge base. We have experimented by choosing different Wikipedia databases (English, Hindi) in enriching document representation. The Keyword vector and the additional vectors are linearly combined for measuring the document similarity. Based on the similarities, clusters are formed separately for English and Hindi documents using Bisecting k-means clustering algorithm. Similarity between the cluster centroids are measured to combine them in order to obtain the final multilingual clusters.

3.1 Document Representation

All the English and Hindi text documents are represented using the classical vector space model [11]. It represents the documents as a vector of keyword-based features following the BOW notation having no ordering information. The values in the vector are TFIDF scores. Instead of maintaining a stopword list for every language, any word that appears in more than 60% of the documents is considered as a stopword. Even after removing stopwords, the document vectors still contain certain unwanted words that lead to noise. Addition of semantic information to these unwanted words might lead to the distortion of original clustering. In order to overcome this problem, we added the semantic information only to a subset of document terms (top-n keywords) that are considered to be important based on their TFIDF scores. We have experimented by considering n values from 40% to 100% with an increment of 10%. Best cluster results were achieved for n=50%. These document vectors form the baseline Keyword vector for our experiments.

3.2 Enriching the Document Representation

To characterize the ever growing content and coverage of Wikipedia, its articles are annotated and categorized. The content of a Wikipedia article is annotated by hyperlinks (references) to other related articles and they denote the "outlinks" for that article. Such outlinks create interlinks between articles. Every article is a description about a single topic or a concept. Equivalent topics (concepts) that are represented with different phrases are grouped together by redirection links. Also, it contains a hierarchical categorization system, in which each article belongs to at least one category. In regard to all the above features, Wikipedia has become a potential resource which can be exploited in enhancing text document clustering.

It was said earlier that the BOW model doesn't consider the semantics of the words. So, two documents with different collections of keywords representing the same topic, may be falsely assigned to different clusters. Hu *et al.* [12] presented an approach in which they have overcome this problem by enriching their document representation with Wikipedia concepts and categories for monolingual document clustering. Along with the Wikipedia concepts and categories, in this paper we have also explored the outlinks and Infobox information. The Infobox of an article contains certain statistical information which represents the most prevalent information in that article. Hence it can be helpful in deciding the topics present in a document and thereby improving the cluster quality.

Wikipedia databases are preprocessed and the title of every Wikipedia article is mapped to its corresponding categories, outlinks, Infobox and the synonyms with the help of Lucene indexer[2]. This mapping process is divided into three steps:

1. For every page of Wikipedia, its title is extracted along with its corresponding outlinks, categories and Infobox information.
2. All the redirections are tracked, as they are considered to be the synonyms for Wikipedia titles.
3. An inverted index is built using Lucene indexer. Lucene builds an index with key-value pairs. Every Wikipedia title forms a pair with each of its corresponding categories, outlinks, Infobox and synonyms. All such pairs are indexed and stored.

A separate index is created for English and Hindi Wikipedia databases.

For every document, its Keyword vector is used for obtaining additional vectors from Wikipedia. For every term in the Keyword vector, either a Wikipedia article with exact title or a redirected article, if present, is fetched. From this article the outlink, category and Infobox terms are extracted to form three additional vectors namely Outlink vector, Category vector and Infobox vector of that document. In all these additional vectors, the values are the TFIDF scores of those terms. As addition of various terms would lead to noise, we have considered only the top-n terms in each of these vectors as we did for the Keyword vector.

E.g. Consider the following document *"It was an exciting match between India and Pakistan. . . . . . . Pakistan prime minister Parvez Musharraf awarded the man of the match to Sachin Tendulkar. . . ."* .

This document is actually about a cricket match. If we enrich the entire document representation and add unwanted information about Parvez Musharraf, prime minister, etc., it degrades the clustering performance. Hence we have considered only the top-n keywords which helped in forming clusters and also reduced the computation time.

3.3 Document Clustering

Document clustering is the automatic grouping of text documents into clusters so that documents within a cluster have high similarity in comparison to one another, but are dissimilar to documents in other clusters. Various clustering approaches (such as Hierarchical clustering, Partitioning clustering, etc.) are available to cluster the enriched documents. Steinbach *et al.* [13] compared different algorithms and concluded that Bisecting k-means performs better than the standard k-means

---

[2] http://lucene.apache.org/java/docs/index.html

and agglomerative hierarchical clustering. We used Bisecting k-means algorithm
[13] for clustering as it combines the strengths of partitional and hierarchical clustering methods by iteratively splitting the biggest cluster using the basic k-means
algorithm. For the evaluation of Bisecting k-means algorithm, we have experimented with fifteen random k values between 30-70 and the average F-score and
Purity values are considered as the final clustering result.

Bisecting k-means algorithm is applied on the additional documents vectors
along with the Keyword vector to obtain separate set of clusters for English and
Hindi documents respectively. We chose the cosine distance, which measures the
similarity of two documents by calculating the cosine of the angle between them.
Each document is represented with four vectors: Keyword vector, Category vector, Outlink vector and Infobox vector.

The similarity between two documents d_i and d_j is defined as follows:

$$sim(d_i, d_j) = sim^{Keyword} + \alpha * sim^{Category} + \beta * sim^{Outlink} + \gamma * sim^{Infobox} \quad (1)$$

Here, $sim(d_i, d_j)$ gives the cosine similarity of the documents d_i, d_j. The sim
is calculated as:

$$sim = cos(v_i, v_j) = (v_i.v_j)/(|v_i| * |v_j|) \quad (2)$$

where v_i and $v_j \in$ {Keyword, Category, Outlink, Infobox} vectors of documents
d_i and d_j respectively. The coefficients α, β and γ indicate the importance of
Category vector, Outlink vector, and Infobox vector in measuring the similarity
between two documents. A similarity measure similar to Eq.(1) was proposed
by Hu *et al.* [12] where the Wikipedia Concepts and Categories were used for
clustering the monolingual documents.

Merging Different Language Clusters: The English and Hindi clusters are
merged based on their centroid similarities. The centroid of a cluster is obtained
by taking the average of all document vectors present in that cluster. In order to
compare English and Hindi clusters, the Hindi cluster centroids are mapped onto
Common Ground Representation (English), the details of which are explained in
Section 3.4. After Common Ground mapping, the similarities of a Hindi cluster
centroid with the centroids of each of the English clusters are measured. The
calculated values are noted in a sorted order. This is repeated for all Hindi
clusters. The two clusters (one English, one Hindi) with the highest similarity are
merged to form a multilingual cluster. This step is repeated with the remaining
set of clusters and finally multilingual clusters are achieved.

3.4 Common Ground Representation

In order to compare multingual cluster centroids, we may need to use the language resources (dictionaries) or language specific tools (such as lemmatizers,
Named Entity Recognizer (NER), Part-of-Speech (POS) tagger). With the inclusion of any language specific tool, re-implementing an approach for a language with fewer resources is a painful process, which limits the extendability
of an approach. So, we have avoided such issues by eliminating the usage of any

such language specific tools in our approach. Instead we have used structured Wikipedia multilingual content and a bilingual dictionary. To cover a broader set of terms, we preferred the Shabdanjali dictionary[3].

Proper nouns play a pivotal role in measuring the similarity between two given documents. Dictionaries, in general, don't cover many proper nouns. Transliteration is highly helpful in identifying the proper nouns, but it requires parallel transliterated (English-to-Hindi) word lists to build even a language-independent statistical transliteration technique [14]. Acquiring such word lists is a hard task when one of the languages is a minority language. Transliteration is included at the cost of reducing the extendability of our approach. As an alternative, we utilized the cross-lingual links that exist in Wikipedia multilingual databases (English and Hindi). The cross lingual links interrelate Wikipedia articles of the different languages. All these articles follow the constraint of sharing identical topics and their titles are verified to be aligned. We availed these alignments in creating a Wiki dictionary to handle proper nouns. This method is language-independent and is easily scalable for other languages. We have mapped all the Hindi cluster centroids onto English using bilingual dictionary and the Wiki dictionary.

Modified Levenshtein Edit Distance Measure: In all the similarities calculated using Equation (1), the terms are compared using the Modified Levenshtein Edit Distance as a string distance measure. In many languages, words appear in several infected forms. For example, in English, the verb 'walk' may appear as 'walked', 'walks', 'walking'. The base form, 'walk', that one might look up in a dictionary, is called the lemma for the word. The terms are usually lemmatized to match the base form of that term. Lemmatizers are available for English and many other European languages. But the lemmatizers support is very limited in the context of Indian languages. So, we have modified the Levenshtein Edit Distance metric to replace the purpose of lemmatizers by adding certain language-independent rules. Henceforth, it can be applied for any language. This modified Levenshtein Edit Distance would help us in matching a word in its infected form with its base form or other in ected forms. The rules are very intuitive and are based on three aspects:

1. Minimum length of the two words
2. Actual Levenshtein distance between the words
3. Length of subset string match, starting from first letter.

No language specific tools were used in this approach. However, to ensure the accuracy, we have worked out alternatives like Wiki dictionary, Modified Levenshtein Edit Distance, etc. which are language-independent.

4   Experimental Evaluation

We have conducted experiments using the FIRE 2010 dataset available for the ad-hoc cross lingual document retrieval task. The data consists of news documents collected from 2004 to 2007 for the English, Hindi, Bengali and Marathi

---

[3] http://ltrc.iiit.ac.in/onlineServices/Dictionaries/Dict_Frame.html

languages from regional news sources. We have worked on only English and Hindi datasets for our experiments. There are 50 query topics represented in each of these languages. We have considered the English and Hindi topic relevant documents to build clusters. To introduce noise, we have added topic irrelevant documents that constitute 10 percent of topic documents. Some topics are represented by 8 or 9 documents whereas others are represented by about 50 documents. There are 1563 documents in the resulting collection, out of which 650 are in English and 913 in Hindi. Cluster quality is evaluated by F-score [13] and Purity [15] measures. F-score combines the information of precision and recall. To compute Purity, each cluster is assigned to the class which is most frequent in the cluster, and then the accuracy of this assignment is measured by counting the number of correctly assigned documents and dividing by total number of documents.

4.1 Wikipedia Data

Wikipedia releases periodic dumps of its data for different languages. We used the latest dump (Sept,'10 release) consisting of 2 million English articles and 55,537 Hindi articles. The data was present in XML format. The required Wikipedia information such as categories, outlinks, Infobox and redirections are extracted and processed for the formation of vectors.

4.2 Discussion

In our experiments, clustering based on Keyword vector is considered as the baseline. Various linear combinations of Keyword, Category, Outlink and Infobox vectors are examined in forming clusters. The cluster quality is determined by F-score and Purity measures. Table 1 displays the results obtained. As mentioned earlier, the coefficients α, β and γ determine the importance of Category, Outlink and Infobox vectors respectively in measuring the similarity between two documents. To determine the α value, we have considered Keyword

Table 1. Clustering schemes based on different combinations of vectors

Notation	F-Score	Purity
Keyword (baseline)	0.532	0.657
Keyword_Category	0.563	0.672
Keyword_Outlinks	**0.572**	0.679
Keyword_Infobox	0.544	0.661
Category_Outlinks	0.351	0.434
Category_Infobox	0.243	0.380
Outlinks_Infobox	0.248	0.405
Keyword_Category_Outlinks	0.567	**0.683**
Keyword_Outlinks_Infobox	0.570	0.678
Keyword_Category_Infobox	0.551	0.665
Category_Outlinks_Infobox	0.312	0.443
Keyword_Category_Outlinks_Infobox	0.569	0.682

and Category vectors to form clusters. Using Eq.(1) experiments are done by varying the α values from 0.0 to 1.0 by 0.1 increment (β and γ are set to 0). The value for which best cluster result is obtained is set as the α value. Similar experiments are repeated to determine β and γ values. In our experiments, it was found that setting $\alpha = 0.1$, $\beta = 0.1$ and $\gamma = 0.3$ yielded good results.

From Table 1, it can be observed that clustering using Wikipedia has performed better than the baseline. Moreover, the outlinks information has proved to perform better than categories followed by Infobox information. Outlinks are the significant informative words in a Wikipedia article which refer (hyperlinks) to other Wikipedia articles. As the outlinks nearly overlap the context of a document, that might have improved the results better than the rest. With the categories, we get generalizations of a Wikipedia topic. Considering the categories, the documents are compared at an abstract level which might have declined the results compared to outlinks. The Infobox provides vital statistical information of a Wikipedia article. Its information is inconsistent across all articles which explains its poor performance when compared with others.

5   Conclusion and Future Work

In this paper, we proposed an approach for enhancing MDC performance by exploiting different features of Wikipedia (such as cross-lingual links, outlinks, categories and Infobox information) and tested it with Bisecting k-means clustering algorithm. Our results showcase the effectiveness of Wikipedia in enhancing MDC performance. The outlinks information has proved to be crucial in improving the results, followed by Categories and Infobox information. The English index has a broader coverage of Wikipedia articles compared to Hindi index and hence the document enrichment is benefited more with English index, this phenomenon is reflected in the results as well. We have avoided use of any language-specific tools in our approach by creating alternatives like Wiki dictionary, Modified Levenshtein Edit Distance, etc. to ensure the accuracy. If bilingual dictionaries are given, this approach is extendable for many other language pairs that are supported by Wikipedia. The system is easy to reproduce and also considers the future growth of information across languages.

We plan to extend the proposed approach, which implements only static clustering to handle the dynamic clustering of multilingual documents. In addition to the aligned titles of Wikipedia articles that have cross lingual links, we further plan to consider the category and Infobox information in building a robust dictionary, eliminating the use of bilingual dictionary and hence achieving a language independent approach. We would also like to consider comparable corpora of different languages to study the applicability of our approach.

References

1. Pirkola, A., Hedlund, T., Keskustalo, H., Järvelin, K.: Dictionary-based cross-language information retrieval: Problems, methods, and research findings. Information Retrieval 4, 209–230 (2001)

2. Silva, J., Mexia, J., Coelho, C., Lopes, G.: A statistical approach for multilingual document clustering and topic extraction form clusters. In: Pliska Studia Mathematica Bulgarica, Seattle, WA, pp. 207–228 (2004)
3. Rauber, A., Dittenbach, M., Merkl, D.: Towards automatic content-based organization of multilingual digital libraries: An english, french, and german view of the russian information agency novosti news. In: Third All-Russian Conference Digital Libraries: Advanced Methods and Technologies, Digital Collections, Petrozavodsk, RCDI (2001)
4. Leftin, L.J.: News blaster russian-english clustering performance analysis. Technical report, Columbia computer science Technical Reports (2003)
5. Romaric, B.M., Mathieu, B., Besançon, R., Fluhr, C.: Multilingual document clusters discovery. In: RIAO, pp. 1–10 (2004)
6. Evans, D.K., Klavans, J.L., McKeown, K.R.: Columbia news blaster: multilingual news summarization on the web. In: HLT-NAACL–Demonstrations 2004: Demonstration Papers at HLT-NAACL 2004, pp. 1–4. Association for Computational Linguistics, Morristown (2004)
7. Steinberger, R., Pouliquen, B., Hagman, J.: Cross-lingual document similarity calculation using the multilingual thesaurus eurovoc. In: Gelbukh, A. (ed.) CICLing 2002. LNCS, vol. 2276, pp. 415–424. Springer, Heidelberg (2002)
8. Steinberger, R., Pouliquen, B., Ignat, C.: Exploiting multilingual nomenclatures and language-independent text features as an interlingua for cross-lingual text analysis applications. In: Proc. of the 4th Slovenian Language Technology Conf., Information Society (2004)
9. Hu, J., Fang, L., Cao, Y., Zeng, H.J., Li, H., Yang, Q., Chen, Z.: Enhancing text clustering by leveraging wikipedia semantics. In: SIGIR 2008: Proceedings of the 31st Annual International ACM SIGIR Conference on Research and Development in Information Retrieval, pp. 179–186. ACM, New York (2008)
10. Wang, P., Domeniconi, C.: Building semantic kernels for text classification using wikipedia. In: KDD 2008: Proceeding of the 14th ACM SIGKDD International Conference on Knowledge Discovery and Data Mining, pp. 713–721. ACM, New York (2008)
11. Salton, G., Wong, A., Yang, C.S.: A vector space model for automatic indexing. Commun. ACM 18, 613–620 (1975)
12. Hu, X., Zhang, X., Lu, C., Park, E.K., Zhou, X.: Exploiting wikipedia as external knowledge for document clustering. In: KDD 2009: Proceedings of the 15th ACM SIGKDD International Conference on Knowledge Discovery and Data Mining, pp. 389–396. ACM, New York (2009)
13. Steinbach, M., Karypis, G., Kumar, V.: A comparison of document clustering techniques. In: TextMining Workshop, KDD (2000)
14. Ganesh, S., Harsha, S., Pingali, P., Varma, V.: Statistical transliteration for cross language information retrieval using hmm alignment and crf. In: Proceedings of 3rd International Joint Conference on Natural Language Processing, IJCNLP, Hyderabad, India (2008)
15. Zhao, Y., Karypis, G.: Criterion functions for document clustering: Experiments and analysis. Technical report, Department of Computer Science, University of Minnesota (2002)

Applying Web Usage Mining for Adaptive Intranet Navigation

Sharhida Zawani Saad and Udo Kruschwitz

School of Computer Science and Electronic Engineering,
University of Essex, Colchester, UK
szsaad@essex.ac.uk

Abstract. Much progress has recently been made in assisting a user in the search process, be it Web search where the big search engines have now all incorporated more interactive features or be it online shopping where customers are commonly recommended items that appear to match the customer's interest. While assisted Web search relies very much on *implicit* information such as the users' search behaviour, recommender systems typically rely on *explicit* information, expressed for example by a customer purchasing an item. Surprisingly little progress has however been made in making *navigation* of a Web site more adaptive. Web sites can be difficult to navigate as they tend to be rather static and a new user has no idea what documents are most relevant to his or her need. We try to assist a new user by exploiting the navigation behaviour of previous users. On a university Web site for example, the target users change constantly. In a company the change might not be that dramatic, nevertheless new employees join the company and others retire. What we propose is to make the Web site more adaptive by introducing links and suggestions to commonly visited pages without changing the actual Web site. We simply add a layer on top of the existing site that makes recommendations regarding links found on the page or pages that are further away but have been typical landing pages whenever a user visited the current Web page. This paper reports on a task-based evaluation that demonstrates that the idea is very effective. Introducing suggestions as outlined above was found to be not just preferred by the users of our study but allowed them also to get to the results more quickly.

Keywords: Web usage mining, adaptive Web sites, evaluation.

1 Motivation

The explosive growth of the Web has contributed to an increasing demand for Web personalization systems. Personalized information technology services have become more and more popular, taking advantage of the knowledge acquired from the analysis of the users' navigational behaviour or usage data. Web usage mining (WUM) aims at discovering interesting patterns of use by analyzing Web usage data. We can explore search patterns with implicit features that exist in the logs of information retrieval and filtering applications [5]. These implicit features

A. Hanbury, A. Rauber, and A.P. de Vries (Eds.): IRFC 2011, LNCS 6653, pp. 118–133, 2011.

can be automatically incorporated into personalization components, without the intervention of any human expert [9].

Unlike personalized search systems, our research aims to work on usage mining techniques to provide a user with customized recommendations, by customizing the contents of a Web site with respect to the needs of a group of users rather than individual users. Let us use a university Web site as an example. Every user of that Web site (be it a student, a member of staff or an external visitor) will have different interests. For the sake of simplicity let us assume that we treat all users of that Web site as part of a single group. While each user has individual interests, we can expect a lot of overlap. A newly registered student who is searching for the teaching timetable is not alone. A lot of other students share this information need and will have searched (and hopefully located) the appropriate documents on the Web site. Hence, a Web site presented according to the entire user community's information access activities is likely to make it easier for new users to find relevant documents quickly. Note, that there is no need to enforce such customization, this can easily be switched off and the Web site appears as normal. The point of the above example is that we are hoping to capture both user and community navigation trails to make the entire Web site adaptive and customize it according to the profile which captures the community or group activities. This allows us for example to exploit a lot of the implicit feedback that those students who leave the institution have left over a number of years which normally remains unused and is lost.

Originally, the aim of Web usage mining has been to support the human decision making process. Using WUM techniques, it is possible to model user behaviour, and therefore, to forecast their future movements [1]. The information mined can subsequently be used in order to personalize the contents of Web pages. Using WUM for personalization has brought promising results, the knowledge discovered through the usage mining process serves as operational knowledge to personalization systems [9]. Realizing the potential of WUM techniques to construct this knowledge, our research aims to provide customized recommendations to the users, as an output from analyzing the users' navigational behaviour or usage data. In our case we are not so much interested in recommendations tailored to individual users but to groups of users.

In order to capture feedback from real users of the adaptive Web sites, we conducted an evaluation to explore the potential of an adaptive Web site. We used an existing university Web site as the baseline system. We then constructed search tasks. Our aim was to be as realistic as possible so we constructed these tasks based on frequently submitted queries recorded on that Web site.

The research questions we tried to answer are as follows:

- Do users find an adaptive Web site useful?
- Do users get to the results quicker using the adaptive Web site?
- Do users find an adaptive Web site easy to use?
- Does the comparison of the two approaches (adaptive versus baseline) show any differences in user satisfaction?

Obviously, the term *adaptive* can be interpreted in many different ways. In the context of this paper we use it to describe the sort of system motivated above, namely a Web site that incorporates recommendations which have been derived from past users' navigation trails.

We will first discuss related work (Section 2) followed by a brief research outline (Section 3). We will then describe the experimental setup (Section 4) followed by a discussion of the results (Section 5). The paper will finish with conclusions and an outlook on future work in Section 6.

2   Related Work

The idea of adaptive Web sites is to automatically improve their organization and presentation by learning from user access patterns [25]. To the best of our knowledge there has been very little progress recently in making Web sites truly adaptive by simply adding links and suggestions to an existing (static) Web site. A lot of work has been reported on mining the users' clickthrough patterns and understanding query intent to improve search result sets, e.g. [13, 2]. However, this is mostly seen in a search context and of not much assistance to a user who accesses and navigates a Web site. Adaptive Web sites can make popular pages more accessible, highlight interesting links, connect related pages, and cluster similar documents together. Perkowitz and Etzioni discuss possible approaches to this task and how to evaluate the community's progress. The focus is either on customization: modifying Web pages/site's presentation in real time to suit the needs of individual users; or optimization: altering the site itself to make navigation easier for all. One user's customization does not apply to other users; there is no sharing or aggregation of information across multiple users, and transformation has the potential to overcome both limitations [27]. Apart from that, index page synthesis is a step towards the long-term goal of change in view: adaptive sites that automatically suggest reorganizations of their contents based on visitor access patterns [26].

We focus on customization: to modify a Web site's presentation to suit the needs of individual or (in our case) groups of users. Compared to related work, the main difference is to adapt its content and presentation based on the group profile for local Web site or intranet access. Apart from that, the system will not be relying on explicit user feedback or working with any webmaster intervention.

Web mining has been proposed as a unifying research area for all methods that apply data mining to Web data [20]. Web mining is traditionally classified into three main categories: Web content mining, Web usage mining, and Web structure mining. Web usage mining aims at discovering interesting patterns of use by analyzing Web usage data [9]. That is the area our research falls into.

Usage patterns extracted from Web data have been applied to a wide range of applications [31]. Different modes of usage or mass user profiles can be discovered using Web usage mining techniques that can automatically extract frequent access patterns from the history of previous user clickstreams stored in Web log files [24]. These profiles can later be harnessed towards personalizing the Web site to the user. Apart from Web usage mining, user profiling techniques can be performed to form a complete customer profile [7]. Web usage mining has been suggested as a new generation of personalization tools, e.g. [32], but note that our research is not going to personalize the *results* provided by search engines, but focuses on a customized presentation of the Web site.

Collaborative filtering (CF) is a related area, it describes the process of filtering or evaluating items through the opinions of other people [30]. Ratings in a collaborative filtering system may be gathered through explicit or implicit means, or both. Collaborative filtering enables the Web to adapt to each individual user's needs. Content-based filtering and collaborative filtering have long been viewed as complementary. Our approach relies on feedback that is entirely implicit and does not need any explicit ratings.

3   Research Outline

The preliminary architecture of our adaptive Web site and some relevant data structures have already been discussed in [29]. The data flow is as follows:

- A user's activity is logged based on the user's search and navigation behaviour (possibly exploiting much more than the search and navigation trails the user leaves behind, in fact it could include any activities the user is happy to share with the logging system such as reading and writing emails, reading documents on the desktop etc).
- Web log analysis techniques are used to trace a user's behaviour or preferences.
- Instead of identifying individual user profiles we collate all logged data to build a *community profile*.
- The Web site is customized by identifying those links on each page present on the Web site that closely *match* the community profile (this matching is done here based on page visits but more sophisticated similarity metrics, e.g. text-based, can be incorporated).

Information retrieval interaction structures have been shown to be useful in exploratory search: a history of queries, documents, search terms, and other objects created or identified during a search session can be used to aid the information seeking process [10]. The data being captured can include the Web site's URL, the title of a Web page, date and time of the activities, and the machine's identification.

For the experiment described here we applied a very simple logging and adaptation process. The adaptive Web site is identical to the existing university Web site apart from suggested links that are highlighted on every page (given that log data has been collected for that page). The three most commonly visited links found on a page are highlighted (by three small stars). When a user hovers over any of the links, the system will further reveal the three most popular links that can be found on the target page. Figure 1 presents an example, three links are highlighted and further suggestions being made for the link *summer schools*.

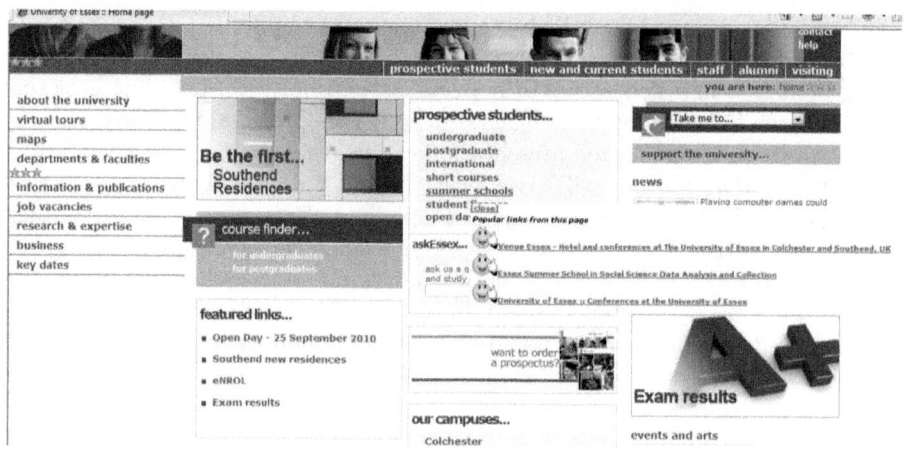

Fig. 1. University home page with embedded recommendations

We bootstrapped the logging tool by recruiting five users (involving postgraduate students and staff) who were willing to perform the same eight tasks as we used in the experiment. These users used the existing Web site to perform the tasks. All their interaction with the existing Web site was logged automatically by the system. These users were not asked to fill in the questionnaires, nor were these users involved in the actual experiments.

4   Experimental Setup

We conducted a task-based evaluation using the Web site of the University of Essex. We used a within-subjects experimental design and in total, 16 subjects participated. Each user had to conduct 8 search tasks and used 2 systems, one being the adaptive Web site, the other one a baseline system. Before starting the experiment, subjects were given a 15 minute introduction to the system.

The two systems can be characterised as follows:

– **System A** is a copy of the existing university Web site and serves as a baseline in this experiment as it is the system that most users will use when trying to locate information on the university Web site.
– **System B** is the adaptive system that adds a layer of commonly followed links on top of *System A* as described above.

The experimental details for our evaluation such as subjects and tasks is based on commonly used standards in task-based evaluations, e.g. [4, 11, 35, 36, 21]. The protocol of the experiments and the time given for each search task are also based on the standard protocol and method suggested by previous interactive search experiments, e.g. [6, 28, 38, 17, 19, 11, 16, 21, 37]. Further explanation on questionnaires is discussed later on.

4.1   Tasks

We constructed a number of tasks to be performed by the users. The tasks were constructed to reflect realistic user needs and were therefore based on some of the most frequently submitted queries derived from the existing intranet query logs [3]. These are the type of tasks we try to target with the adaptive Web site as described in this paper. To avoid potential learning effects and task bias we allocated tasks using a Latin square matrix as explained further down. Furthermore, tasks were constructed based on the brief review guideline suggested by [23], and using real tasks and real users suggested by [8] and [33]. Each user performed four tasks for each system, contributing eight tasks altogether. Examples of the tasks being performed are listed below (the query which the task is based on is shown in parentheses and this query was not part of the task given to the subjects):

- Task 1 (**accommodation**): You have just been accepted for a place at the University of Essex at the Colchester campus. Find information on the residences, accommodation information for new students, contact details and other useful information.
- Task 4 (**short courses**): Find a document that tells you what short courses the University of Essex offers in Business and Management Training.
- Task 7 (**summer school**): Locate a page with useful information about summer school programs and conferences to be held at the University of Essex, and the conference facilities provided by the university.

Users were allocated a maximum of 5 minutes for each task. They were asked to use the system presented to them, either the adaptive Web site or the baseline; to find an answer to the task set. Users were not restricted in any way as to how to use the Web site. They were allowed to either navigate or use the site search engine or a combination of both.

4.2   Data Capture

We used questionnaires and system logging in this experiment. Following TREC Interactive Track guidelines and related work described earlier we used four different questionnaires. The entry questionnaire collected demographic and Internet usage information. A second questionnaire was filled in after every single task a user had conducted to assess the user's perspective of the systems and the tasks. After conducting all tasks on one system a third questionnaire was used to capture the user's perceptions for each of the systems. Finally, after completing all tasks an exit questionnaire was used focusing on a comparison between the two systems.

We logged all user interaction with the systems which included information such as the task time, the title and the links of Web pages visited. Through a detailed analysis of these logs, a comparison of the search behaviour using the adaptive Web site on the one hand and the baseline on the other hand could be carried out.

Table 1. Searcher-by-question matrix

Searcher	System: Questions	System: Questions
1	B:4-7-5-8	A:1-3-2-6
2	A:3-5-7-1	B:8-4-6-2
3	A:1-3-4-6	B:2-8-7-5
4	A:5-2-6-3	B:4-7-1-8
5	B:7-6-2-4	A:3-5-8-1
6	B:8-4-3-2	A:6-1-5-7
7	A:6-1-8-7	B:5-2-4-3
8	B:2-8-1-5	A:7-6-3-4
9	A:4-7-5-8	B:1-3-2-6
10	B:3-5-7-1	A:8-4-6-2
11	B:1-3-4-6	A:2-8-7-5
12	B:5-2-6-3	A:4-7-1-8
13	A:7-6-2-4	B:3-5-8-1
14	A:8-4-3-2	B:6-1-5-7
15	B:6-1-8-7	A:5-2-4-3
16	A:2-8-1-5	B:7-6-3-4

4.3   Questionnaires

A within-subjects laboratory experiment was conducted to compare the systems. This study was conducted in an office in a one-on-one setting. System and task orders were rotated and counterbalanced. Subjects were asked to complete the entry questionnaire. This was followed by a demonstration of the first system. Subjects then used this system to complete four search tasks. After each task, subjects completed the post-search questionnaire. After completing the four tasks, subjects completed the post-system questionnaire. Subjects were then given a demonstration of the second system and asked to complete four more search tasks. After each task, subjects completed the post-search questionnaire. After completing the last four tasks, subjects completed the post-system questionnaire for the second system followed by the exit questionnaire. Each individual experiment lasted approximately 1 hour.

All the questionnaires were constructed based on the TREC-9 Interactive Track guidelines.[1] The set of questions for usability; post-search and post-system, and exit questionnaires were modified based on the measures being used by [38, 21, 34, 11]. This was done in order to map the questions with the intranet tasks and search needs. The entry questionnaire was based on the TREC-9 Interactive Searching Study and the experiment done by [38].

The assignment of subjects to tasks was based on the searcher-by-question Latin square matrix displayed in Table 1. This table contains the mapping of tasks to searchers as proposed in [12].

---

[1] http://www-nlpir.nist.gov/projects/t9i/

5   Results and Analysis

5.1   Subjects

In order to get a good selection of different types of users and to avoid any bias in the selection process we advertised the experiment on campus and selected the first 16 people who replied. Out of the 16 participants 7 were male and 9 female. Their ages ranged overall from 16 to 35 (56.3 percent belongs to 16-25 years, and 43.8 percent belongs to 26-35 years group of age). We had a variety of backgrounds, which included 15 students from different departments and disciplines (including English Language and Literature, Applied Linguistics, Computer Science, Engineering, Banking, Accounting and Finance, Psychology, Philosophy and Politics, International Human Rights and Law), as well as one member of academic staff.

All subjects declared that they use the Internet on a regular basis. The average time subjects have been doing online searching is 8.4 years (13 of them between 6 to 15 years). When asked for their searching behaviour, 12 (or 75%) of the participants selected daily.

Some other interesting statistics are summarized in Table 2 (based on a 5-point Likert scale, where 1 means "none" and 5 means "a great deal"). Note that our users (who we would consider typical target users of the system) tend to have a lot of experience in using Web search systems (mean: 4.44) but little experience with using commercial search engines (mean: 1.94).

Table 2. Subject experience with search systems

Experience	Mean
Searching library catalogues	2.69
Searching commercial online systems	1.94
Searching the Web	4.44
Searching on other systems	2.00

5.2   Average Completion Time

Table 3 gives a picture of the average completion time broken down for each task. We decided to measure the time between presenting the search task to the users and the submission of the result.

Table 3. Average Completion Time (in minutes)

System	Task 1	Task 2	Task 3	Task 4	Task 5	Task 6	Task 7	Task 8
A	3.75	3.50	4.63	3.25	4.25	3.38	3.25	3.25
B	3.13	3.13	2.25	3.13	4.38	3.13	2.50	1.88

Overall, the average time spent on a search task on System A was 3.66 minutes, on System B 2.94 minutes, with statistical difference (p <0.05). This shows that users managed to conduct the search quicker using System B. Further detailed evaluation results are discussed in the following sections.

5.3 Post-search Questionnaire

After finishing each search task a post-search questionnaire had to be filled in. The questions for both systems were the following (using a 5-point Likert scale, where 1 means "not at all" and 5 means "extremely"):

- "Are you familiar with the search topic?"
- "Was it easy to get started on this search?"
- "Was it easy to do the search on this topic?"
- "Are you satisfied with your search results?"
- "Did you have enough time?"
- "Did your previous knowledge help you with your search?"
- "Have you learned anything new about the topic?"

Table 4. Post-search questionnaire (user satisfaction for each task)

System	Task 1	Task 2	Task 3	Task 4	Task 5	Task 6	Task 7	Task 8
A	3.75	4.13	4.14	4.13	3.86	4.25	4.29	3.88
B	4.88	4.63	4.63	4.75	4.13	4.88	4.38	4.38

Table 4 gives a breakdown of the results for the question "Are you satisfied with your search results?" Overall users were more satisfied with the results returned by System B than with System A (with statistical significance, $p < 0.05$, for tasks 1, 2, 3, 4 and 8).

Table 5. Post-search questionnaire (enough time)

System	Task 1	Task 2	Task 3	Task 4	Task 5	Task 6	Task 7	Task 8
A	4.5	4.38	4.43	4.25	4.0	4.38	4.29	4.63
B	4.88	4.63	4.75	4.88	4.5	4.88	4.75	4.75

We had given subjects only 5 minutes to conduct one task and wanted to see if the time allocated was sufficient. Table 5 presents the results for the question "Did you have enough time?" Overall users indicated that they had enough time when using System B as well as System A.

In Table 6 , we give a task-by-task breakdown of some of the properties. Here we do not distinguish between the two systems, because we want to get a picture of what the users' perceptions were about the difficulty of the tasks in general. Note that there seems to be no obvious correlation between familiarity with a topic and the *difficulty* of a task. For example, task 4 is the one users were least familiar with, but they judged 4.06 and 4.25 for *easiness* (to get started with and do the search, respectively), which are higher than some other tasks with much more familiarity.

Table 6. Post-search questions (by task)

Criterion	Task 1	Task 2	Task 3	Task 4	Task 5	Task 6	Task 7	Task 8
Familiarity	3.63	2.69	2.53	1.75	2.0	2.44	1.93	2.69
Start	3.75	4.31	4.33	4.06	3.33	4.0	3.6	4.06
Search	3.81	4.31	4.27	4.25	3.67	4.13	3.73	3.81
Satisfied	4.31	4.38	4.4	4.44	4.0	4.57	4.34	4.13
Time	4.69	4.5	4.6	4.57	4.25	4.63	4.52	4.69

5.4 Post-system Questionnaire

After performing four search tasks on one system a post-system questionnaire had to be filled in. Here we only present the statistics for the multiple choice questions. Later we will discuss any additional comments made by the subjects in more detail. Table 7 gives a breakdown of the results. Two statistically significant results were found; one is users found System B easier to learn to use than System A ($p < 0.0005$), and the other is users found System B more useful in helping them accomplish their search tasks than System A ($p <= 0.0005$). These are important results since we consider the baseline system (System A), i.e. the university Web site, to be very simple and easy to learn to use, since it has been commonly used by all of the participants.

Table 7. Post-system questionnaire

Question	System A	System B
How easy was it to learn to use this information system?	3.75	4.38
How easy was it to use this information system?	4.06	4.31
How well did you understand how to use the information system?	3.88	3.94
How useful was the information system in helping you accomplish your search tasks?	3.81	4.38

The perhaps most interesting finding is users found System B to be more useful in helping them to accomplish the search tasks. This is the main aim of providing an adaptive Web site to the university intranet users.

5.5 Exit Questionnaire

In the exit questionnaire users were asked to answer the question "Which of the two systems did you like the best overall?". Users strongly prefered System B. 12 users prefered System B, 3 prefered System A and 1 found no difference. A large majority of users judged that System B was more helpful in completing tasks than System A, only 2 out of 16 users found System A to be more helpful, and 1 user found no difference. Furthermore, the majority of users also judged that System B was easier to use than the baseline system (10 users found System

B easier, 4 users found System A easier and 2 users found no difference). When looking at the question of which system was easier to learn to use, it is interesting that 8 users found System B easier to learn to use, 6 users found system A easier, and 2 users found no difference in learning to use.

Table 8 summarizes these results. Displayed are the numbers of users who selected each of the choices. The figures in this table confirm the results of the post-system questionnaires in that users found System B much easier to use, and also System B was easier to learn to use, although the first question in the post-system questionnaire (Table 7) would suggest a bigger preference for System B than what users actually expressed in the exit questionnaire. In any case, the fact that users found the adaptive system easier to learn to use is interesting but perhaps not intuitive as it involves additional features that do not necessarily make learning to use the system easier.

Table 9 summarizes the answers users gave in the exit questionnaire concerning the search experience they had in the experiment (where 1 means "not at all" and 5 means "completely").

The main conclusion that we derive from the statistical evidence is that in the given context of intranet search and navigation, users strongly prefer a Web site that offers recommendations as to what pages have been looked at by other users. Our users also managed to conduct the search tasks in less time when being given such recommendations. We can further conclude that users consider such a system to be significantly more useful and easier to use. We can also conclude that the presented adaptive Web site based on a (deliberately simple) "community profile" was generally considered sensible by the Web site users. This offers huge potential in applying a more sophisticated approach that does not just look at frequently visited pages and which builds different adaptive sites based on different user groups (e.g. internal users versus external visitors).

Further experiments need to be conducted though to validate the findings and conclusions because the setup for this initial evaluation involved a relatively small logging history and we tried to answer whether users who conducted a restricted set of specific tasks could be assisted by knowledge acquired from the search history of users who were trying to conduct exactly those tasks.

Table 8. Exit questionnaire (system preference)

Criterion	System A	System B	No difference
More helpful	2	13	1
Easier to learn to use	6	8	2
Easier to use?	4	10	2
Best overall	3	12	1

Table 9. Exit questionnaire (search experience)

Question	Mean
How different did you find the systems from one another?	3.5

5.6   User Feedback

Users were encouraged to leave feedback in addition to answering specific questions. Most users provided comments that would be beneficial for future work. Out of 16 users, 8 users provided comments addressing their perceptions and suggestions on how to improve System B. All 8 preferred System B over System A. One of the comments is *"to keep the suggestions on System B, but they may not pop up unless the user wants them to show"*. One of the users wrote *"I hope that the whole university could change to system B, as it provides the reader with a great deal of help"*. Another user commented *"I like system B more, it is somehow helpful because of the notice using the cursor before you are directed to the right pages"*. One user also mentioned *"The links which appear in system B are very effective, but they should be better organised"*.

Other comments include *"perhaps if the suggested drop down menu would only show if you right clicked or something along that line. This way people can choose to use it or not without being distracted"*, and *"System B makes it easier to browse the university Web site for beginners"*. Apart from that, two other users suggested to improve the search bar and to keep the pages up to date, issues that are more related to the actual Web site.

The user feedback suggests that the simplicity of incorporating suggestions appears to work but that it might also distract people if hovering over a link reveals further suggestions. We have clearly seen the potential of simply exploiting the logged navigation data to improve a Web site. How exactly this is best incorporated needs to be explored in further experiments.

6   Conclusions and Future Work

Our research has addressed the question of whether a Web site can be made more adaptive and easier to navigate by incorporating users' navigation trails without changing the actual Web site. We were looking at a site where each page was simply augmented by recommended links.

To do this we looked at a specific university Web site and evaluated the proposed techniques in a task-based evaluation that was set up to make the evaluation as realistic as possible. We compared the existing Web site against an adaptive site. The results of the evaluation demonstrated that even for the relatively simple approach of just highlighting frequently visted Web pages we can see a clear preference of the adaptive system over the unaltered Web site. The evaluation results are further supported by the fact that it took less time to conduct the tasks when being able to follow recommended links. This is a very strong indication that the outlined idea can be applied successfully on local Web sites such as university sites. However, we would argue that company intranets offer the same potential.

Obviously, despite following all the common evaluation procedures a task-based evaluation as conducted here suffers from a number of shortcomings including the small number of subjects and the fact that a task-based evaluation can only approximate real user experience. A more realistic evaluation will involve setting up an actually adaptive Web site without asking users to conduct manually crafted tasks. This is part of our future work.

There is room for a lot of future research. In the experiment we simply exploited knowledge about visited pages. However, it is easily possible to enrich this log data by logging much more.[2] This is similar to what [15] propose, who suggest that in addition to traditional loggers, data-collection instruments are needed that enrich log data; as such data could normally provide more information-seeking context. Previous evaluation done by [14] used a logging software in order to capture subjects' interaction with all applications including the operating system, Web browsers, and word processors; which were recorded and stored in a protected data file located on the laptop. A user-centered evaluation done by [18] also used a logging software to capture analysts' interactions with systems throughout much of the evaluation.

Another strand of future work is the idea of having different community profiles for different types of users. We used a university Web site as an example where we can distinguish, for example, user communities/groups such as students, staff, external visitors, etc. Having different suggestions for members of individual user groups seems to be a sensible thing. The difficulty will be to allocate users to groups automatically (unless they are signed in). In a company intranet this will be much easier as users are usually grouped into an organisational structure and building different profiles for different groups of users will be more straightforward.

The starting point for this research was the idea that we can add an additional layer on top of an existing Web site. It is of course perfectly reasonable to extend this and insert links that might not originally have been there or to combine this work with alternative ideas of adaptation such as adaptive search, e.g. [22,3].

There are a number of further issues that need to be investigated including the risk that the system might become a self-fulfilling prophecy and the fact that an adaptive system as outlined here might make it harder to reach pages which are less commonly accessed. Furthermore, there will always be a delay in adjusting to small or even significant changes to the Web site.

One major issue not discussed in this paper is the issue of privacy which is paramount when logging personal interaction. Any of the techniques we proposed here allow an easy opt-out for the user but it is an area that needs careful consideration.

In summary, what we presented in this paper addresses issues that come from a variety of research communities including information retrieval, user modelling as well as natural language processing for IR (though this has not been exploited in the experiment presented here). It also offers opportunities for automatically acquiring knowledge which could benefit Semantic Web technologies.

---

[2] We use Jim Jansen's wrapper:
http://faculty.ist.psu.edu/jjansen/academic/wrapper.htm

Acknowledgements

This research was supported by the AutoAdapt[3] research project. AutoAdapt is funded by EPSRC grants EP/F035357/1 and EP/F035705/1.

We would also like to thank the anonymous reviewers for helpful comments and suggestions.

References

1. Baraglia, R., Silvestri, F.: Dynamic personalization of web sites without user intervention. Communications of the ACM 50(2), 63–67 (2007)
2. Bayir, M.A., Toroslu, I.H., Cosar, A., Fidan, G.: Smart miner: A new framework for mining large scale web usage data. In: Proceedings of WWW 2009, pp. 161–170. ACM, New York (2009)
3. Dignum, S., Kruschwitz, U., Fasli, M., Kim, Y., Song, D., Cervino, U., De Roeck, A.: Incorporating Seasonality into Search Suggestions Derived from Intranet Query Logs. In: Proceedings of the IEEE/WIC/ACM International Conferences on Web Intelligence (WI 2010), Toronto, pp. 425–430 (2010)
4. Diriye, A., Blandford, A., Tombros, A.: When is system support effective? In: IIiX 2010, August 18-21, pp. 55–64. ACM, New York (2010)
5. Dumais, S., Joachims, T., Bharat, K., Weigend, A.: Implicit measures of user interests and preferences. In: SIGIR 2003 Workshop Report, pp. 50–54 (2003), SIGIR Forum
6. Dupont, G., Requier, S.A., Adam, S., Lecourtier, Y., Grilheres, B., Brunessaux, S.: A step toward an adaptive composition of query suggestion approaches. In: IIiX 2010, August 18-21, pp. 271–274. ACM, New York (2010)
7. Eirinaki, M., Vazirgiannis, M.: Web mining for web personalization. ACM Transactions on Internet Technology 3(1), 1–27 (2003)
8. Elsweiler, D., Ruthven, I.: Towards task-based personal information management evaluations. In: SIGIR 2007, pp. 23–30. ACM, Amsterdam (2007)
9. Girardi, R., Marinho, L.B.: A domain model of web recommender systems based on usage mining and collaborative filtering. Requirements Eng. 12(1), 23–40 (2007)
10. Golovchinsky, G., Pickens, J.: Interactive information seeking via selective application of contextual knowledge. In: IIiX 2010, August 18-21, pp. 145–154. ACM, New York (2010)
11. Harper, D.J., Kelly, D.: Contextual relevance feedback. In: Information Interaction in Context, pp. 129–137 (2006)
12. Hersh, W.R., Over, P.: Trec-9 interactive track report. In: Proceedings of the Ninth Text Retrieval Conference (TREC-9), pp. 41–50. NIST Special Publication 500-249 (2001)
13. Hu, J., Wang, G., Lochovsky, F., Sun, J.-T., Chen, Z.: Understanding user's query intent with wikipedia. In: Proceedings of WWW 2009, pp. 471–480. ACM, New York (2009)
14. Kelly, D., Belkin, N.J.: Display time as implicit feedback: Understanding task effects. In: SIGIR 2004, pp. 377–383. ACM, Sheffield (2004)
15. Kelly, D., Dumais, S., Pederson, J.O.: Evaluation challenges and directions for information-seeking support systems. Computer 42(3), 60–66 (2009)

---

[3] http://autoadaptproject.org

16. Kelly, D., Fu, X.: Eliciting better information need descriptions from users of information search systems. Information Processing and Management 43(2007), 30–46 (2006)
17. Kelly, D., Harper, D.J., Landau, B.: Questionnaire mode effects in interactive information retrieval experiments. Information Processing and Management 44(2008), 122–141 (2007)
18. Kelly, D., Kantor, P.B., Morse, E.L., Scholtz, J., Sun, Y.: User-centered evaluation of interactive question answering systems. In: Proceedings of the Interactive Question Answering Workshop at HLT-NAACL 2006, pp. 49–56. Association for Computational Linguistics, New York City (2006)
19. Kelly, D., Wacholder, N., Rittman, R., Sun, Y., Kantor, P., Small, S., Strzalkowski, T.: Using interview data to identify evaluation criteria for interactive, analytical question-answering systems. Journal of the American Society for Information Science and Technology 58(7), 1032–1043 (2007)
20. Kosala, R., Blockeel, H.: Web mining research: a survey. SIGKDD Explorations 2(1), 1–15 (2000)
21. Kruschwitz, U., Al-Bakour, H.: Users want more sophisticated search assistants: Results of a task-based evaluation. Journal of the American Society for Information Science and Technology 56(13), 1377–1393 (2005)
22. Kruschwitz, U., Albakour, M.-D., Niu, J., Leveling, J., Nanas, N., Kim, Y., Song, D., Fasli, M., Roeck, A.D.: Moving towards Adaptive Search in Digital Libraries. In: Advanced Language Technologies for Digital Libraries. Springer, Heidelberg (forthcoming, 2011)
23. Kules, B., Capra, R.: Creating exploratory tasks for a faceted search interface. In: Second Workshop on Human-Computer Interaction and Information Retrieval, HCIR 2008 (October 2008)
24. Nasraoui, O., Soliman, M., Saka, E., Badia, A., Germain, R.: A web usage mining framework for mining evolving user profiles in dynamic web sites. IEEE Transactions on Knowledge and Data Engineering 20(2), 202–215 (2008)
25. Perkowitz, M., Etzioni, O.: Adaptive Web Sites: an AI Challenge. Artificial Intelligence 11(1), 246–271 (1997)
26. Perkowitz, M., Etzioni, O.: Adaptive web sites: Conceptual cluster mining. Artificial Intelligence 17(1), 243–273 (1999)
27. Perkowitz, M., Etzioni, O.: Towards adaptive web sites: Conceptual framework and case study. Artificial Intelligence 118(1), 245–275 (2000)
28. Qu, P., Liu, C., Lai, M.: The effect of task type and topic familiarity on information search behaviours. In: IIiX 2010, August 18-21, pp. 371–375. ACM, New York (2010)
29. Saad, S.Z.: Web personalization based on usage mining. In: The 3rd BCS IRSG Symposium on Future Directions in Information Access, FDIA 2009, pp. 15–21 (2009)
30. Schafer, J.B., Frankowski, D., Herlocker, J., Sen, S.: Collaborative Filtering Recommender Systems. In: Brusilovsky, P., Kobsa, A., Nejdl, W. (eds.) Adaptive Web 2007. LNCS, vol. 4321, pp. 291–324. Springer, Heidelberg (2007)
31. Srivastava, J., Cooley, R., Deshpande, M., Tan, P.-N.: Web usage mining: Discovery and applications of usage patterns from web data. SIGKDD Explorations 1(2), 12–23 (2000)
32. Teevan, J., Dumais, S.T., Horvitz, E.: Beyond the commons: Investigating the value of personalizing web search. User Modeling and User-Adapted Interaction 13(1), 311–372 (2005)

33. Wacholder, N., Kelly, D., Kantor, P., Rittman, R., Sun, Y., Bai, B.: A model for quantitative evaluation of an end-to-end question-answering system. Journal of the American Society for Information Science and Technology 58(8), 1082–1099 (2007)
34. Walker, M., Kamm, C., Litman, D.: Towards developing general models of usability with paradise. Natural Language Engineering 6(3), 363–377 (2000)
35. White, R.W., Jose, J.M., Ruthven, I.: An implicit feedback approach for interactive information retrieval. Information Processing and Management 42(2006), 166–190 (2004)
36. White, R.W., Kelly, D.: A study on the effects of personalization and task information on implicit feedback performance. In: Proceedings of CIKM 2006, Arlington, Virginia, USA, pp. 297–306 (2006)
37. White, R.W., Ruthven, I., Jose, J.M.: The use of implicit evidence for relevance feedback in web retrieval. In: Crestani, F., Girolami, M., van Rijsbergen, C.J.K. (eds.) ECIR 2002. LNCS, vol. 2291, pp. 93–109. Springer, Heidelberg (2002)
38. Yuan, X., Belkin, N.J.: Investigating information retrieval support techniques for different information-seeking strategies. Journal of the American Society for Information Science and Technology 61(8), 1543–1563 (2010)

Search Result Caching
in Peer-to-Peer Information Retrieval Networks

Almer S. Tigelaar, Djoerd Hiemstra, and Dolf Trieschnigg

University of Twente
P.O. Box 217, 7500 AE, Enschede, The Netherlands
{tigelaaras,hiemstra,trieschn}@cs.utwente.nl

Abstract. For peer-to-peer web search engines it is important to quickly process queries and return search results. How to keep the perceived latency low is an open challenge. In this paper we explore the solution potential of search result caching in large-scale peer-to-peer information retrieval networks by simulating such networks with increasing levels of realism. We find that a small bounded cache offers performance comparable to an unbounded cache. Furthermore, we explore partially centralised and fully distributed scenarios, and find that in the most realistic distributed case caching can reduce the query load by thirty-three percent. With optimisations this can be boosted to nearly seventy percent.

1 Introduction

In peer-to-peer information retrieval a network of peers provide a search service collaboratively. We define a peer as a computer system connected to the Internet. The term peer refers to the fact that in a peer-to-peer system all peers are considered equal and can both supply and consume resources. In a peer-to-peer network each additional peer adds extra processing capacity and bandwidth in contrast with typical client/server search systems where each additional client puts extra strain on the server. When such a peer-to-peer network has good load balancing properties it can scale up to handle millions of simultaneous peers. However, the performance is strongly affected by how well it can deal with the continuous rapid joining and departing of peers which is called *churn*.

We study peer-to-peer information retrieval systems where each peer contains, and maintains an index over, a subset of all the documents in the system. Since presumably relevant search results can be located at any peer it is difficult to route a query to the right one. This problem is commonly approached using different network topologies and replication of index data [1, 2, 3].

In this paper we explore search result caching. We assume that for each query there is a peer that can provide a set of original search results. If this query is posed often that peer would cripple under the demand for providing this set over and over again. Hence, we propose that each peer that obtains search results for a particular query *caches those results*. The effect is that search results for popular queries can be obtained from many peers: *high availability*, and the load on the peer that provided the original results is reduced: *load balancing*.

A. Hanbury, A. Rauber, and A.P. de Vries (Eds.): IRFC 2011, LNCS 6653, pp. 134–148, 2011.

We define the following research questions:

1. What fraction of queries can be potentially answered from caches?
2. How can the cache hit distribution be characterised?
3. What is the distribution of cached result sets given an unbounded cache?
4. What is the effect of bounding the cache: how does the bound and cache policy affect performance?
5. What optimisations can be applied to make caching more effective?
6. How does churn affect caching?

Most research in peer-to-peer information retrieval focuses on simulating networks of hundreds [4] to thousands [3] of peers. In contrast, our experiments are of a larger scale: using over half a million peers. To our knowledge, there is no previous scientific work that investigates the properties of networks of this size. Our motivation is that large peer-to-peer information retrieval networks deserve more attention because of their real-world potential [5], and that this size is in the range of operational peer-to-peer networks used for other applications [6].

This paper is organised as follows: we discuss related work in Section 2. We explain our experiment set-up in Section 3 and show the results of experiments in sections 4 and 5. Finally, Section 6 concludes the paper.

2   Related Work

Markatos [7] analysed the effectiveness of caching search results for a centralised web search engine combined with a caching web accelerator. His experiments suggest that one out of three queries submitted has already been submitted previously. Cache hit ratios between 25 to 75 percent are possible. He showed that even a small bounded cache (100MB) can be effective, but that the hit ratios still increase slowly when the cache size is increased to several gigabytes. The larger the cache, the less difference the policy for replacing items in the cache makes. He recommends taking into account both access frequency and recency.

Skobeltsyn and Aberer [4] investigated how search result caching can be used in a peer-to-peer information retrieval network. When a peer issues a query it first looks in a distributed meta-index, kept in a distributed hash table, to see if there are peers with cached results for this query. If so, the results are obtained from one of those peers, but if no cached results exist, the query is broadcast through the entire network. The costs of this fallback are $O(n)$ for a network of n peers. In our experiments we do not distribute the meta-index, but focus only on the distributed cache. An additional difference is that they always use query subsumption: obtaining search results for subsets of the terms of the full query. They claim that with subsumption cache hit rates of 98 percent are possible as opposed to 82 percent without. The authors also utilized bounded caches, but do not show the effect of different limits.

Bhattacharjee et al. [8] propose using a special data structure combined with a distributed hash table to efficiently locate cached search result sets stored for particular term intersections. This is particularly helpful in approaches that

store an inverted index with query terms as it reduces the amount of network
traffic necessary for performing list intersections for whole queries. This could be
considered to be bottom-up caching: storing results for individual terms, then
combinations of terms up to the whole query level. Whereas subsumption is
top-down caching: storing results for the whole query, then for combinations of
terms and finally for individual terms.

3   Experiment Set-Up

3.1   Introduction

Our experiments intend to give insight into the *maximum benefits* of caching.
Each experiment has been repeated at least five times, averages are reported,
no differences were observed that exceeded 0.5 percent. We assume that there
are three types of peers: *suppliers* that have their own locally searchable index,
consumers that have queries to issue to the network, and *mixed peers* that have
both. In our experiments the indices themselves do not actually exist and we
assume that for each query a fixed set of pre-merged search results is available.
We also assume that all peers cooperate in caching search result sets.

3.2   Collection

To simulate a network of peers posing queries we use a large search engine query
log [9]. This log consists of over twenty million queries of users recorded over a
time span of three months. Each unique user in the log is a distinct peer in our
experiment for a total of 651,647 peers. We made several adjustments. Firstly,
some queries are censored and appear in the log as a single dash [10]: these
were removed. Secondly, we removed entries by one user in the log that poses
an unusually high number of queries: likely some type of proxy. Furthermore, we
assume that a search session lasts at most one hour. If the exact same query was
recorded multiple times in this time window, these are assumed to be requests for
subsequent search result pages and are used only once in the simulation. Table
1 shows statistics regarding the log. We play back the log in chronological order.
One day in the log, May 17th 2006, is truncated and does not contain data for
the full day. This has consequences for one of our experiments described later.
For clarity: we do not use real search results for the queries in the log. In our
experiments we make the assumption that specific subsets of peers have search
result sets and obtain experimental results by counting hits only.

Table 1. Query log statistics

Users	651,647
Queries (All)	21,082,980
Queries (Unique)	10,092,307

3.3   Tracker

For query routing we introduce the *tracker* that keeps track of which peers cache search result sets for each query. This is inspired by BitTorrent [11]. However, in BitTorrent the tracker is used for locating a specific file: *exact search*. A hash sequence based on a file's contents yields a list of all peers that have an exact copy of that particular file. In contrast, we want to obtain a list of peers which have cached search result sets for a specific free-text query: *approximate search*.

The tracker can be implemented in various ways: as a single dedicated machine, as a group of high capacity machines, as a distributed hash table or by fully replicating a global data index over all peers. Let us first explore if a single machine solution is feasible. The tracker needs to store only queries and mappings to all peers in the network. We can make a rudimentary calculation based on our log: storing IPv6 addresses for all the 650,000 peers would take about 10MB. Storing all the queries in the log, assuming an average query length of 15 Bytes [9, 12], would take about 315 MB. Even including the overhead of data structures we could store this within 1GB. Consider that most desktop machines nowadays have 4GB of main memory and disk space in the range of TeraBytes. However, storage space is not the only aspect to consider, bandwidth is equally important. Assume that the tracker is connected to a 100 Megabit line, which can transfer 12.5 MB per second. The tracker receives queries, 15 Bytes each, and sends out sets of peer addresses, let us say 10 per query: 160 Bytes. This means that a single machine can process 81,920 queries per second. This would work even if 12 percent of the participating peers would query it every second.

In our calculation we have made many idealizations, but it shows that a single machine can support a large peer-to-peer network. Nevertheless, there are three reasons to distribute the tracker. Firstly, a single machine is also a single point of failure: if it becomes unreachable, due to technical malfunction or attacks, the peer-to-peer network is rendered useless. Secondly, a single machine may become a bottleneck even outside its own wrongdoing: for example due to poor bandwidth connections of participating peers. Thirdly, putting all this information in one place opens up possibilities for manipulation.

4   Centralised Experiments

Let us first consider the case where one supplier peer in the system is the only peer that can provide search results. This peer does not pose queries. This scenario provides a baseline which resembles a centralised search system. Calculating the query load is trivial in this case: all 21 million queries *have to be* answered by this single central supplier peer. However, what if the search results provided by the central supplier peer can be cached by the consuming peers? In this scenario the tracker makes the assumption that all queries are initially answered by the central peer. When a consuming peer asks the tracker for advice for a particular query, this peer is registered at the tracker as caching search results for that query. Subsequent requests for that same query are offloaded to caching peers by the tracker. When there are multiple caching peers for a query, one is selected randomly. Furthermore, we assume unbounded caches for now.

Figure 1 shows the number of search results provided by the origin central supplier peer and the summed number of hits on the caches at the consumer peers. It turns out that results for about half of the queries need to be given by the supplier at least once. The other half can be served from the caches of other peers. Caching can reduce the load on the central peer by about 50 percent. This suggests that about half the queries we see are unique. Skobeltsyn and Aberer [4] find that only 18 percent of the queries they use are unique. Perhaps this is because their log is a Wikipedia trace as this is inconsistent with our findings and contradicts observations of large web search engines [13, p. 183]

Caching becomes more effective as more queries flow through the system. This is due to the effect that there are increasingly more repeated queries and less unique queries. So, you always see slightly fewer new queries than queries you have already seen as the number of queries increases. Perhaps there is mild influence of Heap's law at the query level [13, p. 83].

How many results can a peer serve from its local cache and for how many does it have to consult caches at other peers? The local cache hit ratio climbs from around 22 percent for several thousand queries to 39 percent for all queries. These local hits are a result of re-search behaviour [14]. The majority of cache hits, between 61 and 78 percent, is on external peers.

Let us take a closer look at external hits. We define a peer's share ratio as:

$$shareratio = \#cachehits/\#queries \qquad (1)$$

where *cachehits* is the number of external hits on a peer's cache: all cache hits that are not queries posed by the peer itself. *Queries* is the number of queries issued by the peer. A *shareratio* of 0 means that a peer's cache is never used for answering external queries, 1 that a peer answers as many queries as it poses, and above 1 indicates that a peer serves results for more queries than it sends.

Figure 2 shows that about 20 percent of peers does not share anything at all. It turns out that the majority of peers, 68 percent, at least serve results for some queries, whereas only 12 percent, about 80,000 peers, serve results for more queries than they issue.

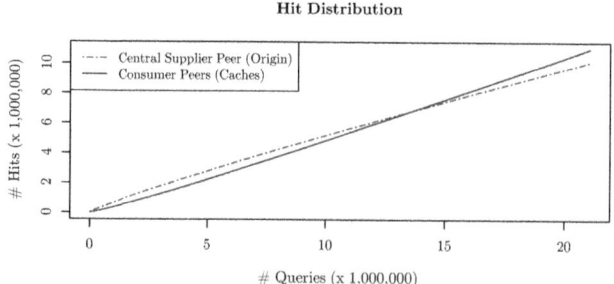

Fig. 1. Distribution of hits when peers perform result caching

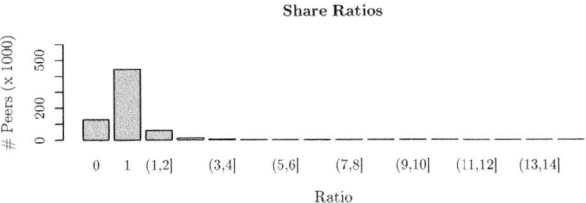

Fig. 2. Observed share ratios

4.1 Required Cache Sizes

So far we have assumed caches of unbounded size. This is not very realistic since machines in a peer-to-peer network have limited resources. Let us try to find out how big a cache we really need. Figure 3 shows the distribution of the number of cached items per peer for the previous experiment. We see that the vast majority of peers, about 225 000, cache between 1 and 5 search result sets. The graph is cut-off after 250 results, but extends to the highest number of cached items seen at a single peer: about 7500.

How much space does it take to store a set of search results? Assume that each set of results consists of 10 items and that each item consists of a URI, a summary and some additional meta-data, taking up 1K of space: 10K per set. Even for the peer with the largest number of cached results this takes only 73MB. However, a cache of 5 items, 50KB, is much more typical. Table 2 gives an overview of storage requirements for various search result set sizes. Most modern personal computers can keep the entire cache in main memory, even with a supporting data structure like a hash table.

Table 2. Cache storage requirements in MegaBytes (MB). Assumes each search result takes up 1KB: 5 results for low, 100 for medium and 7500 for high.

Result Set Size	Low (5)	Medium (100)	High (7500)
10	0.05	1	73
100	0.5	10	730
1000	5	98	7300

4.2 Bounded Caches

As suggested in the previous section: it is possible to use unbounded caches for at least some time. However, it is not very desirable to do so for two reasons. Firstly, if systems run for an extended period of time, the cache has to be bounded somehow since it will run out of space eventually. Secondly, there is no point in keeping around result sets that are not requested any more.

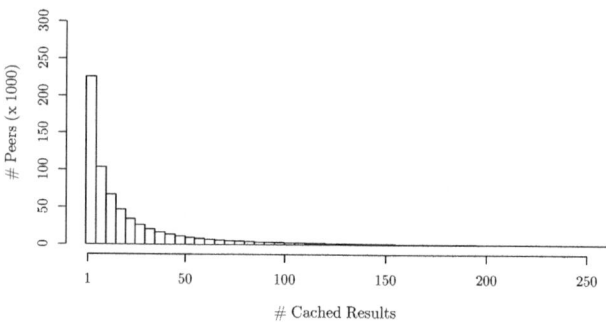

Fig. 3. Observed cache sizes. Each bar represents 5 search results. The horizontal axis extends to 7500. The visible part of the graph covers 99.2 percent of all peers, each peer caches at least one search result.

We want to limit the size of the cache at some maximum number of search result sets to keep. To this end we investigate three different cache policies, with different limits on the cache size. When a new result set has to be inserted in the cache and the cache limit is reached the cache policy comes into play.

The most basic policy when the cache limit is reached is to throw out a random result set, this is called *Random Replacement* (RR) [15]. The advantage of this method is that it requires no additional administration at all. The downside is that we may be throwing away valuable sets from the cache. What is valuable is conventionally expressed using either frequency or recency which provides the motivation for the two other policies tested [7]. In the *Least Frequently Used* (LFU) policy the search result set which was consulted the least amount of times, meaning: which has the least hits, is removed. In *Least Recently Used* (LRU) the search result set that was least recently consulted is removed. In the case of LFU there can be multiple 'least' sets which have the same lowest hit count. If this occurs a random choice is made among those sets.

Figure 4 shows the hit distribution for the baseline unbounded cache and the RR, LFU and LRU caching strategies with various cache limits after running through the entire log. Experiments were conducted with per-peer cache limits of 5, 10, 20, 50 and 100 result sets. We can see that a higher cache limit brings the results closer to the unbounded baseline, which is what we would expect. The most basic policy, Random Removal, performs worst particularly when the cache size is small (L5, L10). However, it performs almost the same as the LFU algorithm for large caches (L50, L100). In fact LFU performs quite poorly across the board. We believe this is caused by the fact that there can be many sets with the same hit count in a cache which degrades LFU to RR. For all cases the LRU policy is clearly superior. Although, the higher the limit, the less it matters what policy is used, also found by [7]. L100/LRU with 10 results per set takes only 1MB of space and achieves 99.1 percent of the performance of unbounded caches.

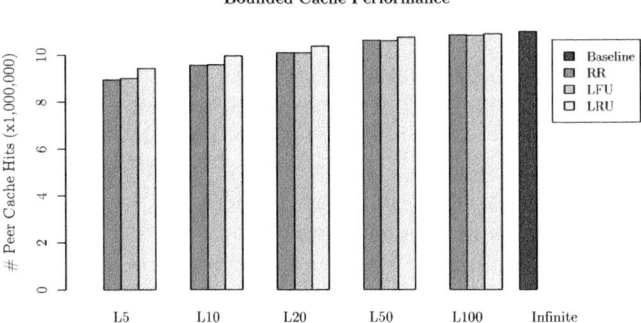

Fig. 4. Bounded cache performance. The total number of queries is 21,082,980. The bars show the amount serviceable from peer caches for various per-peer cache size limits (5, 10, 20, 50 and 100) and strategies (RR, LFU and LRU). The rightmost bar shows the performance with unbounded caches.

4.3   Optimisations

In this section we use unbounded caches and investigate the impact of several optimisations: *stopword removal, reordering, stemming* and *query subsumption*. These techniques map multiple queries that were previously considered distinct to one common query representation. Since the number of representations is lower than the original number of queries, the strain of serving original search results on the central supplier peer is also lower. This capitalizes on the fact that there are cached copies of search result sets around for similar queries.

For stopword removal we remove words from queries that match those in a stopword list used by Apache Lucene consisting of 33 English terms. For re-ordering, the words in the query are alphabetically sorted, for example: from "with MacCutcheon you live happily ever after" to "after ever happily live Mac-Cutcheon with you". The last common technique is stemming, for example from "airplane" and "airplanes" to "airplan". This example also shows the well known drawback of stemming: that of reducing unrelated distinct meanings to the same form. We used the Porter2 English stemming algorithm [16].

We ran experiments with the three described techniques individually and all three combined. The first five rows of Table 3 show the results. We see that without any optimisations the central peer has to serve 47.9 percent of all queries. Applying stopping or re-ordering only marginally improves this by about half a percent. Stemming offers the best improvement: over 1.6 percent. Combining the techniques is quite effective and yields a 3.1 percent improvement in total, which is more than the sum of the individual techniques.

One final technique that is less commonly used is query subsumption [4]. When a full query yields no search results, subsumption breaks the query into multiple subqueries. This process iterates with increasingly smaller subqueries until at least one of these queries yields search results. The subqueries generated are combinations, with no repetition, of the terms in the full query. The length goes down each iteration, starting from $len(query) - 1$ terms to a minimum of 1 term. For example, given a resultless query q of length three: "A B C", we next try the three combinations of length two: "A B", "A C" and "B C". If that yields no results we try all combinations of length one, which are the individual terms "A", "B" and "C". The rationale for iterating top-down, from the whole query to the individual terms, is that longer queries are more specific and are thus expected to yield more specific, higher quality, results. Long queries generate an unwieldy number of possible subqueries. Therefore, we restrict the maximum number of generated combinations at any level to 1000.

In our experiment we evaluate at each iteration whether there is a query that yields at least one search result set. If so: all queries at that same iteration level for which there are cached result sets generate cache hits. Hence, for the example above, if for the full query "A B C" search results are not available, but there is at least one result at the level of individual terms: "A", "B" and "C". The full query can generate 1–3 cache hits: one for each individual term for which a result set is available. This thus causes the total amount of cache hits to increase beyond the number of original queries and simulates the effect of increased query load for merging result sets from multiple peers.

Table 3 shows the results. As mentioned the total amount of cache hits is different: 24 million for subsumption alone, a 13.6 percent increase. Nevertheless, performance improves with 21.4 percent less strain on the central peer. Combining subsumption with the three other techniques further increases the query total to nearly 26 million, but also further decreases the central peer load by 4.2 percent. The trade-off with subsumption is a higher total query load, but a lower load on the central peer. It reduces query-level caching to term-level caching which is known to have higher hit rates [13, p. 183]

All the discussed optimisations decrease precision in favour of higher recall. Hence, the quantity of search results for a particular query goes up, but the quality is likely to go down. Whether such a trade-off is justified depends on how sparse the query space is to begin with. However, for a general search engine, it certainly makes sense to apply some, if not all, of these techniques.

Table 3. Cache hits for various optimizations (x 1,000). Shows what party answers what query as an absolute number and percentage. The first five rows have a total query count of 21 million. The sixth 24 and the seventh 26 million.

	Central		Internal		External	
Baseline	10,092	47.9%	4,237	20.1%	6,754	32.0%
Sto(P)	9,993	47.4%	4,265	20.2%	6,824	32.4%
(R)eorder	9,992	47.4%	4,274	20.3%	6,816	32.3%
(S)tem	9,768	46.3%	4,359	20.7%	6,955	33.0%
P+R+T	9,449	44.8%	4,462	21.2%	7,172	34.0%
S(U)bsumption	6,352	26.5%	7,239	30.2%	10,365	43.3%
P+R+T+U	5,773	22.3%	8,335	32.1%	11,834	45.6%

5   Decentralised Experiments

Now that we have shown the effectiveness of caching for offloading one central peer, we make the scenario more realistic. Instead of a central peer we introduce *n* peers that are *both* supplier and consumer. These mixed peers are chosen at random. They serve search results, pose queries and participate in caching. The remaining peers are merely consumers that can only cache results.

The central hits in the previous sections become hits per supplier in this scenario. We further assume unbounded caches and no optimisations to focus on the differences between the centralised and decentralised case. How does the distribution of search results affect the external cache hit ratios of the supplier peers? We examine two distribution cases:

Single Supplier. For each query there is always only exactly one supplier with unique relevant search results.

Multiple Suppliers. The number of supplier peers that have relevant search results for a query depends on the query popularity. There is always at least one supplier for a query, but the more popular a query the more suppliers there are (up to all *n* suppliers for very popular queries).

For simplicity we assume in both cases that there is only one set of search results per query. In the first case this set is present at exactly one supplier peer. However, the second case is more complicated: among the mixed peers we distribute the search results by considering each peer as a bin covering a range in the query frequency histogram. We assume that for each query there is at least one peer with relevant results. However, if a query is more frequent it can be answered by more peers. The most frequent queries can be served by *all n* suppliers. The distribution of search results is, like the queries themselves, *Zipf* over the suppliers. We believe that this is realistic, since popular queries on the Internet tend to have many search results as well. In this case the random choice is between a variable number *m* of *n* peers that supply search results for a given query. Thus, when the tracker receives a query for which there are multiple possible peers with results it chooses one randomly.

We performed two experiments to examine the influence on query load. The first is based on the single supplier case. The second is based on the multiple suppliers case. For multiple suppliers we first used the query log to determine the popularity of queries and then used this to generate the initial distribution of search results over the suppliers. This distribution is performed by randomly assigning the search results to a fraction of the suppliers depending on the query popularity. Since normally the query popularity can only be approximated, the results represent an ideal outcome. We used $n = 10{,}000$ supplier peers in a network of 651,647 peers in total (about 1.53 percent). This mimics the Internet with a small number of websites and a very large number of surfing clients.

Figure 5 and Table 4 show the results. The number of original search results provided by the suppliers is about five percent higher than in the central peer scenario. This is the combined effect of no explicit offloading of the supplier peers by the tracker, and participation of the suppliers in caching for other queries.

Table 4. Original search results and cache hits (x 1,000). 10,000 peers are suppliers operating in mixed mode

	Single	Multiple
Suppliers (origin)	11,599	12,111
Consumers internal (caches)	3,683	3,930
Consumers external (caches)	5,801	5,042

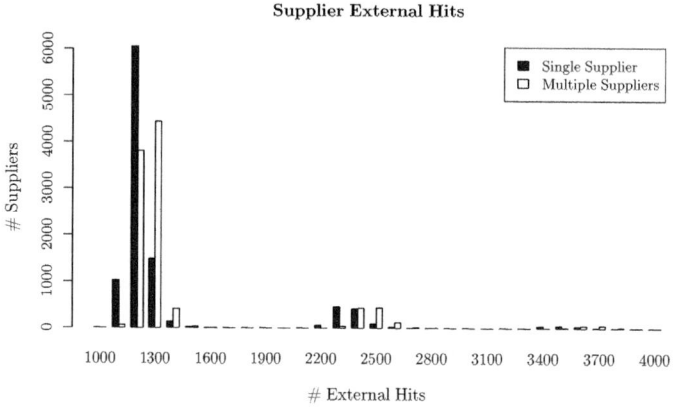

Fig. 5. Supplier external hit distributions (n=10,000 suppliers)

In the second case there is slightly more load on the supplier peers than in the first case: 57 percent versus 55 percent. The hit distribution in Figure 5 is similar even though the underlying assumptions are different. About 87 percent of peers answer between 1000 and 1500 queries. A very small number of peers answers up to about five times that many queries. Differences are found near the low end, which seems somewhat more spread in the single supplier than in the multiple suppliers case. Nevertheless, all these differences are relatively small. The distribution follows a wave-like pattern with increasingly smaller peaks: near 1300, 2500, 3700 and 4900 (not shown). The cause of this is unknown.

5.1   Churn

The experiments thus far have shown the maximum improvements that are attainable with caching. In this section we add one more level of realism: we no longer assume that peers are on-line infinitely. We base this experiment on the single supplier case from the previous section where the search results are uniformly distributed over the suppliers. The query log contains timestamps and we assume if a specific peer has not issued a query for some period of time, its session has ended and its cache is temporarily no longer available. If the same peer

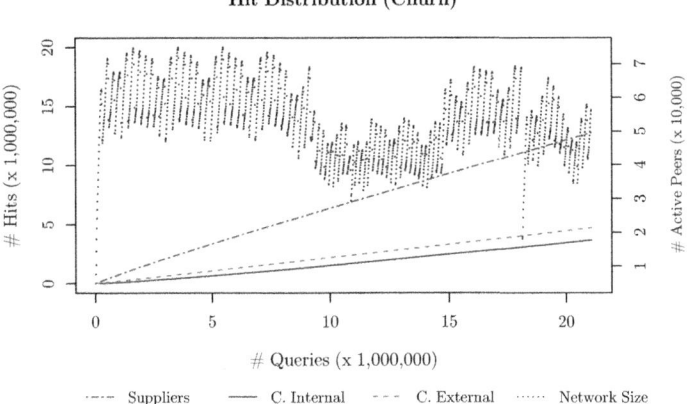

Fig. 6. Distribution of hits under churn conditions (N=651,647 peers)

issues a query later, comes back on-line, its cache becomes available again. This simulates churn in a peer-to-peer network where peers join and depart from the network. We assume the presence of persistent peer identifiers, also used in real-world peer-to-peer systems [17]. All peers, including supplier peers, are subject to churn. For bootstrapping: if there are no suppliers on-line at all, an off-line one is randomly chosen to provide search results.

Assuming that all peers are on-line for a fixed amount of time is unrealistic. Stutzbach and Rejaie [6] show that download session lengths, post-download lingering time and the total up-time of peers in peer-to-peer file sharing networks are best modelled by using *Weibull distributions*. However, our scenario differs from file sharing. An information retrieval session does not end when a search result has been obtained, rather it spans multiple queries over some length of time. Even when a search session ends, the machine itself is usually not immediately turned off or disconnected from the Internet. This leads us to two important factors for estimating how long peers remain joined to the network. Firstly, there should be some reasonable minimum that covers at least a browsing session. Secondly, up-time should be used rather than 'download' session length. As soon as a peer issues its first query we calculate the remaining up-time of that peer in seconds as follows :

$$remaininguptime = 900 + (3600 \cdot 8) \cdot w \tag{2}$$

where w is a random number drawn from a Weibull distribution with $\lambda = 2$ and $k = 1$. The w parameter is usually near 0 and very rarely near 10. The up-time thus spans from at least 15 minutes to at most about 80 hours. About 20 percent of the peers is on-line for longer than one day. This mimics the distribution of up-times as reported in [6], making the assumption the uptime of peers in file sharing systems resembles that of information retrieval systems.

Figure 6 shows the results: the number of origin search results served by suppliers as well as the number of internal and external hits on the caches of consumer peers. We see that the number of supplier hits increases to over 12.75 million: over 1.16 million more compared to the situation with no churn. The majority of this increase can be attributed to a decrease in the number of external cache hits. The dotted cloud shows the size of the peer-to-peer network on the right axis: this is the number of peers that is on-line simultaneously. We can see that this varies somewhere between about 30,000 and 80,000 peers. There is a dip in the graph caused by the earlier described log truncation.

We also ran this distributed experiment with churn with a L100 LRU cache and all optimizations from the previous section enabled: stopping, re-ordering, stemming and query subsumption. This yields a cache hit ratio of 69 percent (for 25,45 million queries) for this most realistic scenario.

6   Conclusion

We conducted several experiments that simulate a large-scale peer-to-peer information retrieval network. Our research questions can be answered as follows:

1. At least 50 percent of the queries can be answered from search result caches in a centralised scenario. For the decentralized case cache hits up to 45 percent are possible.
2. Share ratios, the rate between cache hits and issued queries, are skewed which suggests that additional mechanisms are needed for cache load balancing.
3. The typical cache size is small, with outliers for eagerly consuming peers. Peers that issue a lot of queries also provide lots of cached results.
4. Small bounded caches approach the performance of unbounded caching. The Least Recently Used (LRU) cache replacement policy consistently outperforms the other policies. However, the larger the cache the less the policy matters. If each peer were to keep just 100 cached search result sets the performance is 99.1 percent of unbounded caches.
5. We have shown that stopword removal, stemming and alphabetical term re-ordering can be combined to boost the amount of cache hits by about 3.1 percent. Query subsumption can increase cache hits by 21.4 percent, to nearly 80 percent, but also imposes a higher total query load. All of the optimisation techniques trade search result quality for quantity. However, they all improve the effective usage of caches.
6. Introducing churn reduces the maximum attainable cache hits to 33 percent (-12 percent) without optimizations and 69 percent (-11 percent) with optimizations.

We have shown the potential of caching under increasingly realistic conditions using a single large query log. Caching search results significantly offloads the origin suppliers that provide search results under all considered scenarios using this log. These experiments could be extended by adding extra layers of realism. For example individual search results could be considered instead of fixed search results per query, allowing merging and construction of new search result sets.

We have explored several fundamental caching policies showing that Least Recently Used (LRU) is the best approach for our scenario. However, more advanced policies could be explored that include frequency as a component such as 2Q, LRFU or ARC [7, 18, 15]. These techniques combine advantages of LRU and LFU. Nevertheless, In reality there may be more than just LRU/LFU to take into account. For example queries pertaining to current events for which the relevant search results frequently change. The result sets for such queries should have a short time to live, whereas there are queries for which the search results change very rarely, they could be cached much longer. Making informed decisions about invalidation requires knowledge about the rate of change for particular queries [19]. Perhaps this information, or an estimate thereof, could be made an integral part of the search results, similar to the way in which Domain Name System (DNS) records work. Furthermore, we have assumed that the capacity of the tracker is unbounded. However, a policy similar to what is used to maintain peer caches could be applied there too. This does have the consequence that the tracker looses track of search results which are available, but for which the mappings have been thrown away. Finally, we have not investigated peer selection and result merging, both of which are relevant for real-world systems [5].

Acknowledgements

This paper was created using only Free and Open Source Software. We gratefully acknowledge the support of the Netherlands Organisation for Scientific Research (NWO) under project 639.022.809.

References

[1] Cuenca-Acuna, F.M., Martin, R.P., Nguyen, T.D.: Planetp: Using gossiping to build content addressable peer-to-peer information sharing communities. In: Proceedings of HPDC, Seattle, Washington, US (June 2003)

[2] Suel, T., Mathur, C., Wu, J.w., Zhang, J., Delis, A., Kharrazi, M., Long, X., Shanmugasundaram, K.: Odissea: A peer-to-peer architecture. In: Proceedings of WebDB, San Diego, CA, US, pp. 67–72 (June 2003)

[3] Lu, J., Callan, J.: Full-text federated search of text-based digital libraries in peer-to-peer networks. Information Retrieval 9(4), 477–498 (2006), doi:10.1007/s10791-006-6388-2

[4] Skobeltsyn, G., Aberer, K.: Distributed cache table: efficient query-driven processing of multi-term queries in p2p networks. In: Proceedings of P2PIR, Arlington, Virginia, US, pp. 33–40 (November 2006)

[5] Lu, J.: Full-Text Federated Search in Peer-to-Peer Networks. PhD thesis, Carnegie Mellon University (2007)

[6] Stutzbach, D., Rejaie, R.: Understanding churn in peer-to-peer networks. In: Proceedings of IMC, Rio de Janeiro, BR, pp. 189–202 (October 2006)

[7] Markatos, E.P.: On caching search engine query results. Computer Communications 24(2), 137–143 (2001)

[8] Bhattacharjee, B., Chawathe, S., Gopalakrishnan, V., Keleher, P., Silaghi, B.: Efficient peer-to-peer searches using result-caching. In: Kaashoek, M.F., Stoica, I. (eds.) IPTPS 2003. LNCS, vol. 2735, pp. 225–236. Springer, Heidelberg (2003)

[9] Pass, G., Chowdhury, A., Torgeson, C.: A picture of search. In: Proceedings of InfoScale, Hong Kong, p. 1 (May 2006), doi:10.1145/1146847.1146848

[10] Brenes, D.J., Gayo-Avello, D.: Stratified analysis of aol query log. Information Sciences 179(12), 1844–1858 (2009)

[11] Cohen, B.: Incentives build robustness in bittorrent. In: Proceedings of P2PEcon, Berkeley, CA, US (June 2003)

[12] McNamee, P., Mayfield, J.: Character n-gram tokenization for european language text retrieval. Information Retrieval 7(1), 73–97 (2004), doi:10.1023/b:inrt.0000009441.78971.be

[13] Croft, W.B., Metzler, D., Strohman, T.: Search Engines: Information Retrieval in Practice. Pearson Education, London (2010)

[14] Teevan, J., Adar, E., Jones, R., Potts, M.A.S.: Information re-retrieval. In: Proceedings of SIGIR, Amsterdam, NL, pp. 151–158 (July 2007)

[15] Podlipnig, S., Böszörmenyi, L.: A survey of web cache replacement strategies. ACM Computing Surveys 35(4), 374–398 (2003), doi:10.1145/954339.954341

[16] Porter, M.F.: The english (porter2) stemming algorithm (2001), snowball.tartarus.org/algorithms/english/stemmer.html (January 2011)

[17] Pouwelse, J.A., Garbacki, P., Wang, J., Bakker, A., Yang, J., Iosup, A., Epema, D.H.J., Reinders, M., van Steen, M.R., Sips, H.J.: Tribler: A social-based peer-to-peer system. Concurrency and Computation: Practice and Experience 20(2), 127–138 (2008), doi:10.1002/cpe.1189

[18] Megiddo, N., Modha, D.S.: Arc: A self-tuning, low overhead replacement cache. In: Proceedings of FAST, Berkeley, CA, US, pp. 115–130 (2003)

[19] Blanco, R., Bortnikov, E., Junqueira, F., Lempel, R., Telloli, L., Zaragoza, H.: Caching search engine results over incremental indices. In: Proceedings of SIGIR, Geneva, CH, pp. 82–89 (July 2010), doi:10.1145/1835449.1835466

Author Index

GPSR Compliance

*The European Union's (EU) General Product Safety Regulation (GPSR)
is a set of rules that requires consumer products to be safe and our
obligations to ensure this.*

*If you have any concerns about our products, you can contact us on
ProductSafety@springernature.com*

In case Publisher is established outside the EU, the EU authorized
representative is:

Springer Nature Customer Service Center GmbH
Europaplatz 3
69115 Heidelberg, Germany

Batch number: 09478804

Printed by Printforce, the Netherlands